"[Johnson] knows how to make co[...] easy to follow, and his style is ch[...] deprecating—and laced with those [...] how to find what he calls 'long-decay ideas,' ideas that will stay with us a long time before the last traces vanish from our minds. . . . As he explores his inner world and the mental modules that help to shape it, we begin to feel that we are right in there with him—and we have a new sense of what it means to be human. An entertaining and instructive ride."

—Jonathan Weiner, *The New York Times Book Review*

"A fascinating and graceful tour of the brain . . . With his third book, Johnson . . . has reached that lustrous point in his career where if he's interested in writing about something I'm interested in reading about it. Steven Johnson is using his brain like a grand master in *Mind Wide Open*. The rest of us are lucky to be able to watch."

—Andrew Leonard, Salon.com

"Quirky, captivating tour through the world of modern brain research . . . [Johnson's] thoughts on culture and technology . . . are uncommonly fresh and cogent. . . . The things you've always wondered about your brain . . . are fetchingly elucidated here, presented in a conversational style that makes even forbiddingly difficult concepts seem suddenly graspable. And downright fun. . . . When Johnson concentrates on his lucid and nuanced description of modern neuroscience, of what researchers know and don't know about our brains and why, he's superb. You couldn't have a better guide. You couldn't ask for more enthusiasm or lightly worn expertise."

—*Chicago Tribune*

Praise for Steven Johnson and *Mind Wide Open*

"It's the rare popular-science book that not only gives the reader a gee-whiz glimpse at an emerging field but also offers a guide for incorporating its new insights into one's own worldview. Johnson . . . does just that in his fascinating, engagingly written new survey. . . . [He] weaves disparate strands of brain research and theory smoothly into the narrative . . . which leaves readers' minds more open than they were."

—*Publishers Weekly*

"An enlightening guide . . . an often mind-bending read."

—*Popular Science*

"*Mind Wide Open* is a thought-provoking and engaging book. . . . Johnson's intellectual curiosity, and his enthusiasm for the potential of medical technologies, is infectious."

—*San Francisco Chronicle* (Editor's Pick)

"Thankfully, all the science reads like you're listening to your smartest friend."

—*Men's Health*

"An absorbing narrative . . . fascinating . . . Readers, under the sway of his words and humorous take on life, can't help but marvel along at our brains' abilities and the chemicals that orchestrate them."

—*Los Angeles Times*

"Johnson possesses the rare ability to ground abstract ideas in concrete human realities, while leavening heavy explanations with humor and heart."

—*The Village Voice*

"Spreading a gospel to be curious about one's own mind, Johnson . . . will snare even those unfamiliar with brain science."

—*Booklist*

"Highly readable, smart."

—*The Washington Post*

"Johnson's first-person account of the experiential and neuroscientific aspects of daily life is lucid, illuminating, entertaining, and thought-provoking. You'll find yourself thinking about thinking—while you are thinking—in a whole new way."

—Howard Rheingold, author of *Smart Mobs*

"A fine, broad-based primer on what's happening in brain science."
—*The San Diego Union-Tribune*

"Johnson is an engaging and intelligent guide."
—*Discover* magazine

"My brain was tickled, fascinated, moved, surprised, and above all entertained by Steven Johnson's delightful tour through modern neuroscience."
—John Horgan, author of *The Undiscovered Mind* and *Rational Mysticism*

"Johnson offers a refreshingly personal take on an endlessly fascinating subject."
—*The Guardian* (London)

"Well-researched, informative, and engaging."
—*Cerebrum* magazine

"What good is living in an age of discovery if only a handful of people understand what's being discovered? With this book, Steven Johnson builds an extraordinary bridge between today's trailblazing neuroscientists and the rest of us. His mind-opening and potentially life-changing insight is that virtually anyone can now learn enough about brain chemistry and circuitry to personally explore—and perhaps even reshape—the contours of his or her own mind."
—David Shenk, author of *The Forgetting: Alzheimer's: Portrait of an Epidemic*

ALSO BY STEVEN JOHNSON

*Everything Bad Is Good for You: How Today's
Popular Culture Is Actually Making Us Smarter*

*Emergence: The Connected Lives of Ants, Brains,
Cities, and Software*

*Interface Culture: How New Technology Transforms
the Way We Create and Communicate*

MIND WIDE OPEN

· · · · · · · · · · · · · ·

YOUR BRAIN AND
THE NEUROSCIENCE OF EVERYDAY LIFE

STEVEN JOHNSON

SCRIBNER
New York   London   Toronto   Sydney

SCRIBNER
1230 Avenue of the Americas
New York, NY 10020

First Scribner trade paperback edition 2005

SCRIBNER and design are trademarks of
Macmillan Library Reference USA, Inc., used under license
by Simon & Schuster, the publisher of this work.

For information about special discounts for bulk purchases,
please contact Simon & Schuster Special Sales:
1-800-456-6798 or business@simonandschuster.com.

Designed by Kyoko Watanabe
Text set in Adobe Caslon

Manufactured in the United States of America

1   3   5   7   9   10   8   6   4   2

Library of Congress Cataloging-in-Publication Data
Johnson, Steven.
Mind wide open: your brain and the neuroscience of everyday life/Steven Johnson.
p. cm.
Includes bibliographical references and index.
1. Neurosciences. 2. Neuropsychology. 3. Self-perception. I. Title.
RC341.J648 2004
612.8'2—dc22
2003063308

ISBN 0-7432-4165-7
0-7432-4166-5 (Pbk)

Portions of this book first appeared in *Discover* magazine and *The Nation*.

"Reading the Mind in the Eyes" test and images used by permission of
Simon Baron-Cohen. The test first appeared in the *Journal of Child Psychology
and Psychiatry*, 1997. The "Reading of the Mind in the Eyes" test is based on
photographs from commercial sources. The test itself is used only for research
and is not distributed for commercial profits. Copyright for each individual
photograph cannot be traced from these photo fragments.

For my boys

CONTENTS

MIND WIDE OPEN

... let winged Fancy wander
Through the thought still spread beyond her:
Open wide the mind's cage-door ...

—KEATS

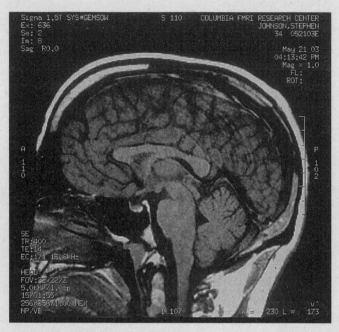

The author's brain, seen through a conventional MRI scan.

Kafka's Room

How pathetically scanty my self-knowledge is compared with, say, my knowledge of my room. . . . There is no such thing as observation of the inner world, as there is of the outer world.

—KAFKA

The idea for this book began with a nervous joke—a handful of nervous jokes, to be precise. A few years ago, thanks to a lucky convergence of events and a long-standing curiosity, I found myself in the office of a biofeedback practitioner, lying on a couch with sensors attached to my palms, fingertips, and forehead. As we talked, the two of us stared into a computer monitor, where a series of numbers flashed on the screen like some kind of low-budget version of the CNBC ticker tape. The numbers documented precisely how much I was sweating and updated several times a second. I've never taken a lie detector test, but something about having a stranger ask me questions while keeping a close eye on my sweat glands put me on edge. And so I started making jokes.

Getting a little tense was partly the point of the exercise. The machine I was attached to was tracking changes in my adrenaline levels, the "fight-or-flight" hormone secreted by the adrenal glands in situations that require a sudden surge of energy. Increased adrenaline can be detected through a number of means: because the hormone diverts blood from the extremes of the body to the core, drops in temperature at the extremities often suggest a release of adrenaline (hence the sensors on my fingertips). Sweating is also a telltale sign of heightened adrenaline levels. Because damp skin conducts electricity more effectively than dry skin, the electrodes on my palms could track how much I was sweating by monitoring changes in conductivity over time.

Biofeedback systems are designed to give you a new kind of control over your body and mind by making physiological changes visible in a new way. After a few sessions, biofeedback users learn to "drive" their adrenaline levels up or down almost as though they were deciding to lift a finger or bend a knee. The brain, of course, is constantly adjusting adrenaline levels anyway—it's just that you're not usually aware of the process other than as a background sense of increased energy or calm.

For the first five minutes of the session, my adrenaline levels remained at the midpoint of the scrolling chart, bouncing around ever so slightly, but with no real pronounced variation. And then something in the situation—I can't remember now what it was—caused me to make an offhand joke. We both chuckled at my remark and then noticed that a huge spike had appeared on the monitor. Making the joke had triggered a surge of adrenaline in me. Or was it the reverse? Perhaps the rise in adrenaline was me mentally revving the engines before launching my joke into the environment. Whatever the causal chain, my joke-telling and my adrenaline levels were locked in some kind of chemical embrace.

The extent of that link became clear at the end of our session, when the therapist handed me a printout of my adrenaline levels plotted over our thirty-minute encounter. It was, simply put, a timeline of my attempts at humor: a flat line interrupted by five or six dramatic spikes. I looked at that paper and thought: I've caught a glimpse of *me* here, viewed from an angle that I've never experienced before. I'd known for many years that I had a tendency to crack jokes compulsively in certain social situations, particularly in situations where the formality of the setting made humor a riskier bet. But I'd never thought about those jokes as triggering a chemical reaction in my own head. Suddenly, they seemed less like casual attempts at humor and more like a drug addict's hungering for a new fix.

I knew those adrenaline surges were just the tip of the iceberg. The creation and appreciation of humor is a remarkably complex neurological event, involving many parts of the brain and a host of chemical messengers. Doctors at the University of California Medical School, for example, recently located a small region near the front of the left brain that appears to trigger the feeling of mirth; while treating a sixteen-year-old epileptic patient, they applied a tiny jolt of electric current to the area, which caused the patient to find humor in whatever she happened to be looking at. This wasn't merely a physical reflex of laughter: things genuinely seemed funny to her when the region was stimulated. ("You guys are just so funny—standing around," she told her startled doctors.) Laughter itself involves a complex array of muscle actions, and there is increasing evidence that it triggers the release of small amounts of endorphins, the brain's natural painkillers. (The next time you visit a comedy club, think "opium den.") But making jokes in conversation also requires a subtle sense of one's audience, a feel for their sense of humor and state of mind. Such outer-directed imagination is itself governed by another part of the brain, a part believed to be damaged in autistics and that accounts for their strained social interactions.

This is what came to my mind as I thought about my nervous

jokes on the biofeedback practitioner's couch: that with each of those jokes somewhere in my head there was an elaborate electro-chemical ballet unfolding, one that had been evolving since my first smile, or before. And now I had glimpsed a subsection of that inner performance as it happened. I found myself wondering how many of these little chemical subroutines are running in my brain on any given day? At any given moment? And what would it tell me about myself if I could see them, the way I could see those adrenaline spikes on the printout?

And so biofeedback started me on my quest. I set out to track down as many charts, real-time displays, and 3-D models of my mental life as I could find. I talked to some of the world's leading neuroscientists, asking them the question I'd been asking myself: "How had understanding the brain changed the way they thought about themselves?" I also found technology startups and armchair enthusiasts who had embraced brain science as a tool for self-exploration. It was a propitious time to make this journey. Over the past three decades, science has given us extraordinary glimpses of the brain's inner geography, illuminating the amazing extent to which different tasks activate clearly defined regions: recognizing the face of a loved one, or planning a grocery list, or stringing together a sentence. Thus far, these new scientific tools have been employed mostly to observe people who have suffered neurological damage and to assess the mental maps shared by all human brains. But brains are like fingerprints—each of us possesses a unique neurological topography. We now have the technology in place to picture that inner landscape, in itself as it really is. These are tools, in other words, for exploring our individual minds, with all their quirkiness and inimitability. These are tools for capturing who we are, on the level of synapses and neurotransmitters and brain waves. Every human brain is capable of generating different patterns of electrical and chemical activity. The promise of these new tools involves being able to figure out what *your* pattern looks like. And then figuring out what that pattern tells you about yourself.

It's likely that you've thought about the patterns of your own brain's wiring before. The general movement of popular psychology over the past century has been one from deeply figurative descriptions of mental traits toward greater physiological specificity: the movement, in a sense, from Oedipus to the neuron. Adrenaline itself has entered our everyday lexicon, as has the notion of our body administering quick chemical fixes purely for pleasure: we do things, we say, for the adrenaline rush, or the endorphin high. Radio ads now tout various wonder drugs' ability to alter our neurotransmitter profiles as though they were selling dandruff shampoo. If you've read *Listening to Prozac,* you've probably met a person who seemed depressed and thought: *hmm, very low serotonin.* But such responses are just hunches about our inner physiological states, and crude ones at that. There are dozens of so-called information molecules in your body—neurotransmitters, hormones, peptides—each playing a key role in your shifting emotional response to external events, triggering everything from the nurturing instinct in mothers to the agitated surge of a panic attack. Could tools that measure the minute-by-minute levels of those substances in your body and brain teach you something about your own emotional toolbox? Could they help you make sense of your dreams, or your phobias? We've learned to track our mood changes with a statistician's exactitude, to explore our childhood memories, to keep our minds alert with exercise. But your moods and memories and perceptions are themselves derived from electrochemical activity in your brain. What could you learn about yourself if you could catch a glimpse of that activity directly? If you could see what your brain looked like when it was remembering a long-forgotten childhood experience, or listening to a favorite song, or conceiving a good idea?

Brain-imaging tools are miracles of modern science, but they are not the only channels to your mind's inner life. Simply possessing a more informed understanding of your brain's internal architecture can change the way you think about yourself. Part of such a process involves separating out mental routines that you typically experience in unison. If you know nothing about what's actually happening in your head, the neurological activity you experience is invisible: it's just you being yourself. But the more you learn about the brain's architecture, the more you recognize that what happens in your head is more like an orchestra than a soloist, with dozens of players contributing to the overall mix. You can hear the symphony as a unified wash of sound, but you can also distinguish the trombones from the timpani, the violins from the cellos. To come to a comparable understanding of your own head, you don't need a million-dollar imaging machine. You just need to learn something about the brain's components and their typical patterns of activation. Sometimes those components come in the form of specialized brain regions; sometimes they come in the form of chemicals, like serotonin. Invariably, a certain mood that strikes you will contain a mix of both, the result of both neurochemical release and predictable activity in specific regions of your brain.

As you learn to detect these brain components, you start to recognize how much multitasking is really going on in your own head. You realize that the emotion you feel isn't simply a reaction to the world at that moment, but rather something closer to a drug, with a strange life of its own. There's what we used to call a "rational" you and an "emotional" you, and the two aren't always in sync. Brain science has now given us more accurate descriptions of these two sides of a personality, mapped onto specific regions of the brain. Instead of "rational" and "emotional," today we have the "neocortical" you and the "limbic" you.

Consider this situation, which you've probably encountered

many times before. You're in a perfectly good mood, having a conversation with a friend or colleague. You're not particularly aware of your emotional state, but it's purring along behind the scenes, making your dialogue free and unencumbered. And then your friend makes a passing reference to something unsettling, maybe a little stressful. Not earth-shattering, not immediately life-jeopardizing, but stressful nonetheless. Maybe he's alluded to some upcoming corporate retreat you haven't been invited to, or a tax deadline you'd forgotten about. Whatever it is, the news triggers a falling sensation in your body; you feel deflated and on edge.

And then your friend says something that surprises or distracts you, and the depressing news flies out of your working memory, replaced by some other thought. At this moment, something uncanny happens in your head, not unlike the feeling of déjà vu. You feel the stress in your body and your head, but you can't remember what triggered it in the first place. The feeling has been separated from the thought. Or put another way, you've lost the thought, but the feeling keeps on churning. Normally in this type of a situation you end up rewinding the tape of the conversation in your head—What were we just talking about?—and you locate the original item after a few seconds, at which point your mental state seems to snap back into place, just like the feeling of déjà vu lifting and linear time reinstating itself. You're still stressed, but at least you know the reason why.

Discontinuities occur like this because your conscious, second-by-second processing of a verbal conversation happens in one part of your brain, while your emotional evaluations happen somewhere else. Most of your immediate focus on generating and comprehending spoken words takes place, broadly speaking, in the prefrontal lobes of the neocortex, the most evolutionarily modern part of the brain. (Two small regions are particularly crucial: Broca's and Wernicke's areas, the former largely focused on creating speech, the latter on processing incoming words.) But the emotions largely issue forth from areas located below the cortex, the region often called the "limbic system," while some of their bodily effects are triggered one layer below the limbic system, in the brain stem that lies at the top of your spinal column. The activity in the prefrontal lobes consists mostly of the flash of neurons talking to each other in a very small region of your head, while the limbic system starts a cascade of events that lead to the release of chemicals that travel throughout the body, including one called "cortisol" that is responsible for much of the physical damage caused by long-term stress.

So when you hear that stress-inducing sentence, two reactions go off in your head: your language centers and working memory decode the meaning and put it front and center in your consciousness; and a subcortical system triggers the stress response, releasing cortisol and other chemicals throughout your brain and body. The two systems operate at fundamentally different speeds, the prefrontal activity unfolding on the level of microseconds and the stress system on the level of seconds or even minutes. That's why the two can get out of sync with one another. You think of something stressful and just as quickly forget about it. The prefrontal lobes can move that fast. But your emotional systems lag behind— there's still cortisol floating in your bloodstream thirty seconds after the news vanishes from your working memory. And so the feeling stays alive in you.

The question is: for that moment of disconnect, what exactly is in charge here? Your frontal lobes or your limbic system? And which one should you trust?

Brain science books sometimes suffer from a recurrent problem, one with no small measure of irony. The subject matter of a book about the human brain is, by definition, as close to home as you get.

(These books are being read by human brains, after all.) But the deeper you delve into the details of brain anatomy, the higher the ratio of Latinate to English words becomes, and before long the lay reader is struggling to keep track of names like the "cingulate cortex" and the "nucleus accumbens." Some books try to scale this learning curve by starting off with a crash course in neuroanatomy. My approach is different: we'll start instead with a brain in action—feeling fear, laughing at a joke, coming up with a good idea—and tease out the underlying mechanisms as we go.

I've also tried to limit the terminology needed to read this book: a half dozen chemicals, a half dozen brain regions, and a rudimentary understanding of the way neurons communicate. It is one of my fundamental assumptions that you can get something useful out of neuroscience with this level of mastery. (For the aficionados and the extracurious, I've included more detailed explanations in the endnotes.) The brain contains multitudes, as Whitman said in another context, but you don't need to memorize them all to be a better user of your brain. If you know the landmarks, you can get your bearings. And when you're navigating a space as complicated as your own brain, getting your bearings can make all the difference.

If you've read a little about the brain over the past decade, you've no doubt encountered two topics that have dominated the public discussion of brain science. The first has to do with explaining consciousness, what the neuroscientist Antonio Damasio calls "the feeling of what happens." The second has to do with the field of evolutionary psychology, which argues that our brains contain a kind of mental toolbox selected over millions of years of evolution to help our ancestors survive and reproduce in challenging environments. Consciousness and evolution are each fascinating avenues for exploration, but this book will try to sidestep both, in slightly different ways.

Let's start with consciousness. Imagine you're seeing the face of a loved one after a long time apart, and feeling the pleasurable emotions triggered by that sight. We know a great deal about the path of incoming visual stimuli, shuttling information about the light bouncing off the contours of the face from your optic nerve to the sensory cortex. We know that this information resonates with memory storage systems controlled by the hippocampus, helping you remember details about your loved one. We also know quite a bit about the chemicals released in your brain that conjure up the feeling of emotional warmth. Thanks both to modern imaging technologies and studies of patients with localized brain damage, we can describe with truly remarkable precision the neurological ballet performed in your head when you gaze at the face of a child or spouse. But our scientific vision grows foggier when we try to explain how those patterns of neurochemical activity somehow create your first-person experience of that gaze: the "faceness" of your loved one's face, the "emotionness" of the emotional feeling. Consciousness theorists call these properties "qualia": the brain's representation of both the external world and the body's internal state—the taste of red wine, the look of light shimmering on water, the feeling of sudden fear hijacking your body.

It seems preposterous at first, but there is a real question as to why we need qualia at all. We could theoretically have evolved brains capable of the entire range of human mental responses—processing internal and external stimuli, evaluating situations as either emotionally positive or negative, executing long-term plans—without actually *feeling* any of these processes. We'd be like robots or zombies, indistinguishable from normal humans from the outside but empty on the inside. So the question becomes: how did this strange property of mind come about? The brain is ultimately just a big lump of atoms strung together in a particular configuration, no different in this sense from a teakettle or a crown of broccoli. Presumably the teakettle and the broccoli aren't conscious of themselves or their environment, so why should we be?

To simplify almost to the point of parody, there are four competing answers to that question on today's consciousness stage. The first is that the broccoli and the teakettle *are* conscious in some unimaginably different way from how we are. In other words, qualia is a property of matter itself, and the human brain is simply the most advanced qualia recording apparatus yet evolved. The second answer is that something unique exists in the configuration of cells that makes consciousness happen in brains and not in broccoli, though the nature of that something is a matter of great debate. The third answer implicates a mystery substance not yet understood by science—quantum behavior, perhaps, or some kind of spiritual life force—that turns a bunch of interconnected cells into a feeling brain. The fourth is the trick answer, proposing that one of the properties of consciousness is that it can't explain itself, and so we'll never get to the bottom of qualia no matter how scientifically and technologically adept we become.

These are all mesmerizing possibilities, even if they do tend to induce a kind of existential vertigo (or make you a little squeamish the next time you drop a piece of broccoli into a pot of boiling water). I wouldn't be at all surprised if one of the many theories of consciousness proposed in the past decade turns out to be largely correct. But science is very far from a consensus on this question right now, and I suspect it will remain in that state for the foreseeable future.

And so in this book, I've made it a matter of policy to avoid the question of consciousness as often as possible. Running away from the problem of qualia turns out to be a relatively healthy strategy, because there's a huge number of interesting and productive things that you can say about the brain without tackling the question of why consciousness feels the way it does. Think about my biofeedback session and my joke-telling adrenaline fix. Getting even that brief glimpse of my brain's chemical feedback system taught me something new about my personality and my conversational habits, and sharpened my awareness of the way making jokes changed my internal mood. (And explained why I sometimes had a tendency to make jokes inappropriately.) But despite these insights, I have no idea whatsoever why an adrenaline rush *feels* the way it does. I can describe its edgy uplift, compare it to the effects of exogenous drugs like caffeine, predict the ways it will change my subsequent behavior. But I can't tell you where the qualia of adrenaline comes from. It would be nice to know, of course, but fortunately it's not the only kind of knowledge that neuroscience can impart to us.

Then there's the evolutionary psychology debate, which runs parallel to—and is often indistinguishable from—the question of nature and nurture. Are our mental faculties simply the product of evolved genes, or are they shaped by the circumstances of our upbringing? Unlike the mysteries of consciousness, this question has a clear, and I believe convincing, answer: they're both. We are a mix of nature and nurture through and through, and it's precisely the interplay between evolved tools and cultural experience that creates the richness of the human condition.

In this book, I discuss some of the properties of the brain in terms of evolution, because a Darwinian perspective can sometimes illuminate features that might otherwise be shrouded in darkness, or help us understand drives and habits of mind that are unduly powerful or hard to shake. In chapter four, for instance, we'll look more closely at the brain science of laughter, and part of that analysis will touch on why laughter evolved in the first place, which in turn helps us understand something new about when and why we laugh in everyday life. (It has much less to do with humor than you might think.)

So evolutionary explanations will not be entirely absent from the chapters ahead, but neither will they be front and center. You can be agnostic about—or downright hostile toward—the premise of the evolved brain and still gain something from modern brain science, because on a basic level, the languages of nature and nurture are written in the same ink. My brain, for instance, may be releasing adrenaline with each successful punch line because millions of years of evolution endowed me with DNA that wired it that way. Or it may be that some unique set of circumstances from my childhood influenced that circuit in my brain. Most likely, of course, it's a bit of both: adrenaline release during laughter may be a common human trait, just a little exaggerated in my case. But whatever the original cause, the wiring is there in my head, releasing its adrenaline like some kind of neurochemical Old Faithful. It's fascinating to speculate whether a specific trait came from your ancestors or your fifth-grade teacher, but you don't need to have a convincing answer to learn about the inner life of your brain.

When public conversation turns to the way our biology shapes our behavior, we often encounter a quick denunciation of the entire premise: someone will claim that talking about minds in biological or Darwinian terms is "biological determinism," a highbrow, sanitized version of the old horrors of racism, eugenics, and social Darwinism. For the most part, these fears are unfounded. Evolutionary psychology addresses the shared characteristics of the human species, what unites us all irrespective of race or culture—exactly the opposite of what a race-based inquiry into our biological roots would attempt to discover.

Of course, the one place in which the evolutionary psychologists have in fact emphasized differences over commonalities is the fraught world of the sexes. Because so much of natural selection is predicated on reproductive success or failure, and because men and women have such different biological stakes in the act of reproduction, and because the sexual divide has been evolving for hundreds of millions of years, and not hundreds of thousands—it is inevitable that natural selection would craft slightly different toolboxes for each sex. Viewed with modern imaging technologies, men's and women's brains are nearly as distinct from each other as their bodies are. They have reliably different amounts of neurons and gray matter; some areas linked with sexuality and aggression are larger in men than in women; the left and right hemispheres are more tightly integrated in women than in men. And of course, those brains—and the bodies they are attached to—are partially shaped by two totally different kinds of hormones, the androgens and estrogens, which play a key role both in development and adult life experiences. Men and women are most certainly not from Mars and Venus, but it is entirely fair to say that they are on different drugs. A world in which the sexes were mentally indistinguishable might be a less conflict-ridden world, though also a little duller. But the truth is it is not the world we inhabit. Writing a book about brain science without describing some of these differences would be an exercise in bad faith, emphasizing politics over science in a way that does injustice to both.

In the past few decades, a certain type of science story has become commonplace in the media. You've probably encountered dozens of renditions of it: scientists announce that they have uncovered the roots of a particular human psychological attribute. The two standard variations of this story are the brain scanning version and the evolutionary psychology version. In the former, scientists pick some trait or behavior—a craving for sugar, say—and use a brain-imaging device to scan someone while they're experiencing that craving. The part of the brain that lights up during the scan—the dorsal striatum, in this case—is identified as the "craving center" of the brain, and before long a press release is being drafted.

The evolutionary psychology version of the same story follows a different path. Instead of locating neurological roots, the scientists discover historical roots: the evolutionary history of why one trait came to be selected. This is a more speculative science, but a powerful one nonetheless. It takes an explanatory approach, not just a descriptive one, trying to answer the ultimate question of why we are the way we are. So the evolutionary psychologists explain that we have sugar cravings because carbohydrates were rare on the savannahs of Africa where the modern human brain evolved. A rule of thumb that was adaptive in one environment (if you happen to find sugar, eat as much of it as you can) turns out to be maladaptive in an environment where Coca-Cola is practically in the water supply.

These two stories are intriguing ones, and there's much to be learned from both approaches. But neither story tells you something about your own present-tense experience that you don't know already. You're already familiar with your sugar cravings, and while it's nice to learn about their origins, knowing the role of the dorsal striatum won't help much the next time you're salivating over that Mars bar. If science is going to tell you something useful about your brain, it has to go beyond simply explaining the roots of some familiar mental phenomenon. Your brain is filled with a lively cast of characters sharing space inside your cranium, and while it's interesting to find out their exact addresses, that information is ultimately unsatisfying. Call it the "neuromap fallacy." If neuro science turns out to be mostly good at telling us the location of the "food craving center," or the "jealousy center," then it will be of limited relevance to ordinary people seeking a new kind of self-awareness—because learning where jealousy lives in your head doesn't make you understand the emotion any more clearly. Those neuromaps will be of great interest to scientists, of course, and doctors. But to the layperson, they'll be little more than trivia.

The best that the brain sciences offer comes in the form of genuine insights, insights in both senses of the word: a looking within and a new way of understanding. To that end, I have applied a test of sorts to the stories I've assembled for this book. I call it the "long-decay" test—as with a sound wave that takes an extended time to trail off into silence (or a radioactive material with a long half-life). There are insights about the brain that prompt a quick burst of recognition—"So that's where the food craving comes from!"—and then just as quickly fade in the mind. These insights fail the long-decay test—they don't stick with you in any profound way. To pass the test, the insight has to reverberate for weeks or months after you've first encountered it; it has to pop up in conversation or in moments of self-reflection; it may even change your behavior based on what it teaches you about yourself. Long-decay ideas transform as much as they inform.

For the most part, the long-decay ideas I've assembled here have direct relevance to ordinary minds, minds untroubled by the extreme conditions profiled in so much of the scientific literature: amnesia, Parkinson's, Alzheimer's, manic-depression, the many forms of aphasia. The most powerful theories of mind have always had something useful to contribute to generally healthy minds and not just troubled ones. Freud developed his theories partially by analyzing the debilitating disorders of hysterics and schizophrenics, but psychoanalysis ultimately attracted such a large audience because you didn't need to be mentally ill to find something useful in it. You could explore your Oedipal complex and analyze your dreams even if you weren't worried about your sanity. I believe modern neuroscience deserves to be seen the same way: as relevant to the healthy as it is to the ill, as relevant to those of us wrestling with the small triumphs and tragedies of everyday life as it is to those battling more forbidding demons.

Enough disclaimers. I've tried to write what follows not as a polemic or a broadside, but as a kind of appreciation. Think of the way an art historian or a musicologist can help you discern new qualities in a great painting or symphony; your perception widens when you look through their eyes or listen with their ears. Brain experts can help us do the same with our own mental life. Under their tutelage, we start noticing reflexes and patterns hitherto invisible to us. Knowing something about the brain's mechanics— and particularly *your* brain's mechanics—widens your own self-awareness as powerfully as any therapy or meditation or drug. Brain science has become an avenue for introspection, a way of bridging the physiological reality of your brain with the mental life you already inhabit. The science and technology today are no longer limited to telling us how *the* mind works. They also have something to say about how *your* mind works.

Unlike so many technoscientific advances, the brain sciences and their imaging technologies are, almost by definition, a kind of mirror. They capture what our brains are doing and reflect that information back to us. You gaze into the glass, and the reflection says to you, "Here is your brain." This book is the story of my journey into that mirror.

1

Mind Sight

"He that has eyes to see and ears to hear may convince himself that no mortal can keep a secret. If his lips are silent, he chatters with his fingertips; betrayal oozes out of him at every pore."

—FREUD

I'm gazing into a pair of eyes, scanning the arch of the brow, the hooded lids, trying to gauge whether they're signaling defiance or panic. Just a pair of eyes—no mouth or torso, no hand gestures or vocal inflections. All I have to go on is a rectangular photo of two eyes staring at me from a computer screen. When I've made my judgment—it's defiance, after all—another set pops on the screen, and I start my examination all over again.

This reverse eye exam is part of an ingenious psychological test devised by the British psychologist Simon Baron-Cohen. The test presents you with thirty-six different sets of eyes, some crinkled in mirth, others gazing off to the horizon deep in thought. Below each image are four adjectives, such as:

despondent
preoccupied
cautious
regretful

Or:

skeptical
anticipating
accusing
contemplative

It's your job to choose the adjective that best fits the image. Is that raised eyebrow a sign of doubt? Or is it rebuke? The eyes them-selves are a demographic mix: some weathered and ancient, others accented with mascara and eyeliner. The subtlety of the expressions is astonishing; as I scroll from image to image, I'm seeing the human eye with a fresh perspective, feeling a newfound amazement at its communicative range.

This test, though, is not ultimately about the eye's capacity to signal emotion. It's about something just as impressive, and just as easily overlooked: the brain's ability to read those signals, to peer into the inner landscape of *another* mind, while relying only on the most transient of cues. You won't find exam questions like these on the IQ test, or the SATs, but the mental skills being measured here are as essential as any in our cognitive toolbox. It turns out that one of the human brain's greatest evolutionary achievements is its abil-ity to model the mental events occurring in other brains.

Chances are you've had an experience roughly like this: you're at a social gathering with colleagues or peers—say it's an office holi-day party—and you run into a coworker with whom you have an unspoken rivalry. It's one of those relationships that is chummy on the surface, but right beneath there's a competitive energy that neither side acknowledges. When you first encounter your colleague, there's the usual pleasant banter, but before long he's confessed to you that something has gone wrong with his career trajectory: either he's lost a big account at work or the fellowship didn't come through or the last batch of short stories got rejected. Whatever it is, it's bad news. It's the sort of news that a friend should perhaps greet with a concerned, doleful expression, which is exactly the expression that you deliberately contort your face into as he delivers the news.

The trouble is, you're only a friend on the surface. Below the surface, you're a rival, and a rival wants to grin at this news, wants to relish the *schadenfreude*. And so for a split second, as you're hearing the fateful syllables roll off his tongue, his tone foreshadowing his disappointment before the sentence is even complete, you let out the slightest hint of a grin.

And then an intricate dance begins. As your face wraps itself up in dutiful concern, you detect a flash of something in *his* face, a momentary startle that says, "Were you just smiling right there?" Perhaps his eyes suddenly lock on to your pupils, or he pauses in midsentence as though something has distracted him. In your mind, an interior closed-captioning emerges: "Did he see that grin?" As you offer your condolences, you can't help wondering if your words sound cruel rather than comforting. "Is he thinking that I'm faking all this sympathy? Maybe I should tone it down a notch just in case."

The silent duet of those two internal monologues should be familiar to you, even if you're the sort of person who never, ever gloats at another's downfall. (Henry James made a literary career out of documenting these subtle interactions.) It needn't be a Cheshire cat grin that provokes the interior monologues: imagine a conversation between two potential lovers, in which one worries that a facial expression has betrayed his love before he has summoned the courage to make a formal declaration. Sometimes the closed-captioning can overshadow the main dialogue, which can make for stilted conversation, with each participant second-guessing the other's thoughts.

This silent conversation—a passing grin, a sudden look of recognition, a lurking question about another's motivation—comes so naturally to us that most of the time we're not even aware that we are locked into such a complex exchange. The internal duet comes naturally because it relies on parts of the brain that specialize in precisely this kind of social interaction. Neuroscientists refer to this phenomenon as "mindreading"—not in the ESP sense, but rather in the more prosaic, but no less impressive, sense of building an educated guess about what someone else is thinking. Mindreading is literally part of our nature. We do it more effortlessly, and with more nuance, than any other species on the planet. We construct working hypotheses about what's going on in other people's heads almost as readily as we convert oxygen into carbon dioxide.

Because mindreading is part of our nature, we don't bother to teach it in schools or test our aptitude for it in placement exams. But it is a skill like any other, a skill that is unevenly distributed throughout the general population. Some people are deft mindreaders, picking up subtle intonational shifts and adjusting their response with imperceptible ease. Others mindread with the subtlety of a Mack truck, constantly second-guessing themselves or interrogating their conversational partners. Some are simply "mindblind," shut off entirely from other people's internal monologues.

Even though we don't teach this particular skill in school, and we barely have a vocabulary to describe it, our mindreading abilities play a key role in our work and relationship successes, our sense of humor, our social ease. But to understand these consequences, you have to stop taking the internal duet for granted. You have to slow it down, explore its underlying processes, recognize the duet for the marvel that it is.

Our growing appreciation for the art of mindreading was accelerated in the late 1990s by the discovery of "mirror neurons" in the brains of monkeys, neurons that fire both when a monkey does a particular task—grabbing a branch, for example—and when the monkey sees another monkey do that same task, suggesting that the brain is designed to draw analogies between our own mental and physical states and those of other individuals. At the same time, researchers explored the premise that autistic people suffer from a kind of mindblindness, preventing them from building hypotheses about others' internal monologues. In related studies, evolutionary psychologists began to think about the Darwinian rewards of mindreading in a social species, examining chimp populations for signs of comparable internal duets. Yet other scientists speculated on the connection between mirror neurons and the origins of language, since all forms of communication presuppose a working model of the object you're attempting to communicate with. For language to evolve, humans needed a viable theory about the minds of other people—otherwise, they'd just be talking to themselves.

Let's now go back to that silent duet at the office party, to the moment that half-concealed grin leaks out of the side of your mouth before you can replace it with the look of sympathy. What's happening here? Most of the time you walk around with the assumption that you're the boss of you, that you have a unified self that controls your actions in a relatively straightforward way. But your telltale grin challenges most of our assumptions about this selfhood, because at that moment at the office party, you are trying your hardest to do the exact opposite of smiling; you're trying to look concerned and upset, full of compassion. But your mouth wants to smile. Whose mouth is it anyway?

The answer is that your mouth has several masters, and some of them are brain subsystems that regulate emotional states. Smiling at times of genuine pleasure is not a learned behavior; every recorded culture on the planet represents the internal mental state of happiness with a smile. Deaf-blind children start smiling on the exact same developmental timetable as children who can see and hear. Cultures certainly differ in their assumptions about what makes people happy, as the popularity of frog's legs and Steven Seagal movies in France will attest. And cultures also differ in their production of fake smiles, as in the beaming "bye-bye nows" of American flight attendants. But genuine happiness—whatever the details of its origin—expresses itself as a smile in all normal homo sapiens.

Ironically, the forced smile of the flight attendant demonstrates just how innate the smiling reflex really is. A century and a half ago, the French neurologist Duchenne de Boulogne began studying the muscular underpinnings of people's facial expressions, using the then-state-of-the-art technologies of photography and electricity. Duchenne photographed his subjects in various emotional states, and tried to automatically simulate their expressions by activating specific muscles with a small jolt of electric current. (The images from Duchenne's experiments look like something from a Nine Inch Nails video.) In 1862, he published his findings in a volume titled *Mechanism of the Human Physiognomy,* which Darwin drew upon extensively ten years later in his best-selling *The Expressions of the Emotions in Man and Animals.* But Duchenne's research soon fell into oblivion, only to be discovered more than a century later by the University of California at San Francisco psychologist Paul Ekman, now generally considered to be the world's leading expert on facial expressions.

The most widely cited discovery in Duchenne's work involved

smiling. Using his crude tools, Duchenne established that genuine smiles and fake smiles utilize completely distinct ensembles of facial muscles—most visible in the eyes, which crinkle in real smiles but remain unchanged in the faux ones. (As a tribute to his long-neglected forebear, Ekman began referring to the genuine article as a "Duchenne smile.") The muscle that controls eye-smiling is called the *orbicularis oculi,* and its activation has proved to be a reliable indicator of internal happiness or mirth. Modern brain scans show that pleasure centers in the brain light up in sync with the *orbicularis oculi,* but show no activity during fake smiles created with the mouth alone. The next time you want to know if your beaming waiter truly wants you to have a nice day, check out the outer edges of his eyebrows; if they don't dip slightly when he smiles, he's faking it.

Duchenne's insights into the muscular underpinnings of the smile make it easier to detect counterfeit good cheer, but they also teach us a more important lesson about selfhood and the emotions. Duchenne smiles are not willed deliberately into existence. You can consciously paint a fake smile on your face, but a real one erupts through a process that your conscious mind controls only in part. This is demonstrated most vividly in studies of stroke victims who suffer from a disturbing condition known as central facial paralysis, which prevents them from voluntarily moving either the left or right side of their face, depending on the location of the neurological damage. When these individuals are asked to smile or laugh on command, they produce lopsided grins: one side of the mouth curls up, the other remains frozen. But when they're told a joke or they're tickled, full smiles animate their face.

This is why the smile has more than one master: sometimes it is triggered by the emotional systems, other times by areas that control voluntary facial movement. (Of course, depending on the brain region, the smile will differ slightly in its expression.) So that inadvertent grin that slips out at the news of your rival's misfortune? It's the result of two brain systems vying for control of the same face. The part of the brain that controls voluntary muscle movement—called the motor cortex—sends a command instructing the face to appear sympathetic. But your emotional system is requesting a toothy grin. Your face can't satisfy both requests at the same time, so what results is a little bit of both: a grin that swiftly morphs into an expression of worried sincerity.

And herein lies lesson one of that office party encounter: your brain is not a general-purpose computer with one unified central processor. It is an assemblage of competing subsystems—sometimes called "modules"—specialized for particular tasks. Most of the time, we only notice these modules when their goals are out of sync. When they work together, they coalesce into a unified sense of self. The idea of multiple selfhood is not, strictly speaking, a discovery of the brain sciences. There's a long tradition of artists and philosophers documenting how fragmented we are below the surface, most notably in the modernist writers that pried open the psyche a century ago. Here's Virginia Woolf describing the struggle between the two models of self in *Mrs. Dalloway:*

> How many million times she had seen her face, and always with the same imperceptible contraction! She pursed her lips when she looked in the glass. It was to give her face point. That was her self—pointed; dartlike; definite. That was her self when some effort, some call on her to be her self, drew the parts together, she alone knew how different, how incompatible and composed so for the world only into one centre, one diamond, one woman who sat in her drawing-room and made a meeting-point . . .

Freud famously envisioned the psyche as a battleground among three competing forces: id, superego, and ego. The modern understanding of the brain shatters that earlier vision into dozens of component parts, some specializing in core survival tasks, such as heartbeat regulation and the fight-or-flight instinct, others focused on more prosaic skills, such as face recognition. Your personality is, in a real sense, the aggregate of the differing strengths of each of these modules—as they have been shaped both by nature and nurture, by your genes and by your lived experience. In other words: *you are the sum of your modules.*

If the modular nature of the mind is often hidden to us, how can we see behind the curtain of the unified self and catch a glimpse of those interacting components? Several avenues are available to us. There are the studies of pathological cases popularized by books such as Oliver Sacks's *The Man Who Mistook His Wife for a Hat,* in which we detect the existence of modules through patients who have suffered targeted brain damage that takes out one or two modules but leaves the rest of the brain functioning normally. Or we can experience the modularity of the brain more directly by taking drugs that throw a monkey wrench into its machinery, causing individual modules to take on a new autonomy (which is why people on drugs often feel as though they hear voices). Or you can gaze inside your brain directly, using today's brain-imaging technologies.

Another more entertaining way into the modular mind is through the back door of illusions and various tricks of the mind. Optical illusions help reveal modules by triggering conflicts between different submodules in the visual system: modules for distinguishing between background and foreground, recognizing borders between objects, or locating objects in 3-D space. Remember the childhood game of spinning in place and then stopping quickly to feel the spinning continue? In this game, as you turn, objects in the room pass by you in a counterclockwise direction. But when you stop, you feel a sense of vertigo, and the room seems to be spinning around you in the reverse direction, as though you were standing at the motionless center of a merry-go-round. Why does the room seem to spin after you've stopped moving? And why does it appear to spin in the other direction?

This staple of early childhood play reveals the brain's modular approach to detecting motion. The part of the brain that evaluates whether you're moving relies on two primary sources: information from the visual field and information from the fluid sloshing around in your inner ear. Most of the time, those two lieutenants concur in their assessments to their commander, but when you stop suddenly after spinning clockwise, the liquid in your inner ear continues to move around for a few seconds more, while your vision responds instantly to the cessation of movement. So the haptic centers of the brain are taking in conflicting data: the inner ear reports you're still moving, while the eyes report that you're at rest. The only way the brain can resolve this conflict is to assume that both reports are correct: you *are* still spinning, but it doesn't seem that way because the world around you is spinning right along with you. The illusion of the world rotating is actually a brilliant on-the-fly interpretation that your brain makes to reconcile the conflicting data it receives. It's not the correct interpretation, of course, but it's a revealing one.

Module disagreement is not a bad way of describing the ultimate cause behind that inadvertent grin at the office party: part of your brain wants to smile, and part of it wants to show sympathy. The result is a kind of "slip of the face": the mouth and eyes betraying an emotion that the social self wants suppressed. The lesson here is that the control structures between modules often matter as much as the strength or weakness of each module itself. The brain is a network, and the way that each node in that network communicates with other nodes is an essential part of its higher-level properties. Even among the macrostructures of the brain, the connections made are as important as the individual structures themselves. One notable difference between male and female neuroanatomy is the communication channel that connects the left and right hemispheres, called the corpus callosum, which is much larger in women than in men. We now believe that this increased connectivity enables women to do a better job than men at reconciling the sometimes conflicting interpretations offered up by each hemisphere.

Some people are good at suppressing grins, while others are lousy at it. Some modules are better at overriding other modules; some are more submissive. Understood in the broadest sense, the process of growing up can be seen as the slow subjugation of emotional centers—such as the amygdala, which plays an essential role in fear responses—by the more recently evolved regions of the brain located in the prefrontal cortex that control voluntary actions, long-term planning, and other higher functions. Infants are born with relatively well-developed amygdalas, which is why they're so good at being frightened right out of the gate. But their prefrontal regions take most of childhood to mature.

So not only is the mind a network of distinct modules, but those modules sometimes compete with each other. The brain's modular system cannot be imagined as a neurological report card, with a B+ for face recognition and a failing grade for mindreading. This is because the modules interact with each other, sometimes inhibiting, sometimes amplifying, sometimes translating or interpreting in novel ways. The brain is much more like an ecosystem than a list of stable personality traits, with modules simultaneously competing and relying on each other. Hence lesson two: *It's a jungle in there.*

So if we now understand something about that renegade grin, what can we say about its detection? The silent duet of mindreading begins in your colleague's brain when he first thinks to himself, midsentence, that you might be quietly celebrating his bad news. It's fitting that the telltale sign is the crinkling of your eyes, as your *orbicularis oculi* betrays your inner state. Mindreading is in many ways a kind of eye-reading—we learn a great deal about the content of other people's thoughts by watching their eyes. Eyes are essential to building what brain scientists call a "theory of other minds."

The connection between mindreading and eye-reading begins early in child development—so early, in fact, that it is unlikely to be the product of learned behavior. In their first year, most children will become adept at something called "gaze monitoring": they see you looking off toward the corner of the room; they turn and look in that direction; then they check back to make sure the two of you are looking at the same thing. Because we do it so well, gaze monitoring doesn't seem like much of an accomplishment, but it requires an elaborate understanding of the human visual apparatus, too elaborate to be purely the product of cultural learning.

Think about what's implied in gaze monitoring. First, you have to understand that people have their own perceptions of the world, distinct from yours. Second, some of those perceptions flow into their mind through their eyes. Third, you can determine the objects people perceive by drawing a straight line from the black circles in the middle of their eyes outward. Fourth, when those black circles shift, that means the gaze has shifted to another object. Consequently, if you want to know what another person is perceiving, you follow the movement of those black circles, and then shift your own gaze toward the object they're focused on.

If the gaze-monitoring skill were purely a learned behavior, it would take a month of school and a four-year-old's brain to master it. Infants can barely be taught how to use a spoon, much less how to track retinal movements and deduce inner mental states. They can't *learn* gaze monitoring, but they do it nonetheless—because their brains contain a cheat sheet of sorts that prepares them for the underlying principles of gaze monitoring, a kind of psychological physics: people have minds; people's minds perceive different things; part of that perception happens through the eyes; if you want to know what someone's thinking, look at his eyes. These biological cues start early in life: one study found that two-month-old infants were more likely to stare at the eyes than at any other part of the face.

As we grow older, we scrutinize people's eyes for subtler cues: not just what they're looking at, but what they're thinking and feeling. Because our emotional systems are wired directly to our facial muscles, à la the Duchenne smile, we often get accurate portraits of other people's moods just by scanning their eyes or the corners of their mouth. As our office party exchange shows, sometimes that portrait gives a more accurate testimony than people's verbal descriptions of their moods. Who are you going to believe—me or my lying eyes?

Gaze monitoring and emotional expression recognition are two of the fundamental mindreading systems, but we also use other tricks. We monitor speech intonation carefully for emotional nuance. We put ourselves into other people's mental shoes—what the cognitive scientists call the "simulation theory" of mindreading, according to which your brain is effectively running a mini-simulation of someone else's to anticipate how the other person might feel.

Your brain runs all these routines any time you interact with other people. It takes careful training, or massive distraction, to stop your mind from inferring other people's mental states as you talk to them. Mindreading is a background process that feeds into our foreground processes; we're aware of the insights it gives us but usually not aware of how we're actually getting that information, and how good we are at extracting it.

The sophistication of our mindreading skills is part of our heritage as social primates; our biology contains cheat sheets for building theories about other minds because our brains evolved—and continue to evolve—in complex social environments where being able to outfox or cooperate with your fellow humans was essential to survival. So just as some animals evolved nervous systems that were adapted for sudden movement or sonar, our brains grew increasingly sophisticated at modeling the behavior of other brains. An entire host of neurological systems revolve around the expectation that you will spend much of your life managing social relationships of one sort or another. Your brain is wired to expect an environment with oxygen, gravity, and light. It's also wired to expect an environment populated by other brains. Hence lesson three: *Deep down, we're all extroverts.*

We're all extroverts, except those of us whose brains have developed without the normal mindreading systems. There are dozens of neurological disorders that compromise social skills, but few are more common than the family of conditions that we generally call "autism."

Autistic people possess many skills lacking in the normal population: they often have nearly photographic memories and astonishing mathematical abilities. Their ease with mechanical systems, including computers, can be extraordinary. But autism impairs social skills dramatically. While autistic people can usually learn and communicate using language, there is something missing in their exchanges with other people, some strange distance in their social demeanor. They seem emotionally remote, disconnected.

Many experts now believe that this distance derives from a distinct neurological condition: autistics are mindreading-impaired. The social distance associated with autism is a vivid example of the brain's modular nature: autistics generally have above-average IQs, and their general logic skills are impeccable. But they lack *social* intelligence, particularly the ability to make on-the-fly assessments of other people's inner thoughts. Autistic people *do* have to go to school to read facial expressions—learning to intuit another person's mood is at least as challenging for them as learning to read is for the rest of us. When you're engaged in conversation, you don't think to yourself, "Aha! His right eyebrow just crinkled up. He must be happy." You just sense that there's a happy expression on his face. But autistics have to perform precisely that kind of deliberate analysis, memorizing which expressions are associated with which emotions and then studying people's faces actively as they talk, looking for signs. One of the early predictors of autism in toddlers is an inability to perform gaze monitoring. It's as though autistics are born without the social physics that the rest of us possess innately, as though they were mindblind.

Simon Baron-Cohen believes that the symptoms of autism exist on a continuum: while some people clearly suffer from extreme cases, millions suffer only from minor cases of mindblindness. (Because autism is ten times more likely to develop in boys than girls, Baron-Cohen has argued that the disorder should be considered simply an extreme version of the male brain's tendencies, rather than a disconnected aberration.) The history of mathematics and physics is populated by borderline autistics: people with great number skills but limited social grace. We all know bright people who perform poorly in social situations, seem disengaged in conversation, or fail to pick up on our emotional cues. Even if you're a particularly astute mindreader, you probably have your own "autistic moments" in passing, when you're conducting a conversation on autopilot, lost in your own internal monologue. If you spend enough time with the literature, you can't help dividing up your friends and colleagues into the talented mindreaders and the mind-dyslexics. You start evaluating your own prowess as you engage with other people. Mindreading becomes a part of your basic vocabulary for evaluating yourself and others: some people have a sharp sense of humor, some are quick learners, some are good mindreaders.

If autism exists on a continuum, then it's possible to locate yourself on that continuum. You can take a simple test called the Autism Spectrum Quotient that Baron-Cohen and his colleagues created—answer fifty questions about yourself on a Web page, and a simple program spits out a number between 1 and 32. The higher the number, the closer you are to autism. (The median result is 16.4.) It's not exactly hard science because it relies on self-evaluation and the questions themselves are relatively broad. But if you trust your ability to assess the general areas of your personality, the test provides a rough sketch of your autism quotient (otherwise known as "AQ").

The questions are phrased as statements with which you can "definitely agree," "slightly agree," "slightly disagree," or "definitely disagree."

> I frequently find that I don't know how to keep a
> conversation going.
> I find it easy to "read between the lines" when someone
> is talking to me.
> I usually concentrate more on the whole picture, rather
> than on the small details.
> I am not very good at remembering phone numbers.
> I don't usually notice small changes in a situation or a
> person's appearance.

If you've read something about autism, or the theory of other minds, these questions will seem predictable enough. When I took the test—if you must know, I scored a 15, just slightly less autistic than average—I flipped through the questions with a kind of jaded awareness: here's the facial expression question, here's the number memory question. It was only when I went back and reviewed the exam that I realized my familiarity with the topic had blinded me to something fascinating about the test itself.

Think about those last two statements: "I am not very good at remembering phone numbers" and "I don't usually notice small changes in a situation or a person's appearance." Now, if you come to the test knowing something about autism, you'll instantly deposit those two statements on opposite ends of the AQ spectrum. An autistic person, you'll think, will be good at remembering phone numbers and bad at noticing small changes in someone's appearance. But if you don't know anything about autism, if you're just coming to the test with a commonsense understanding of human psychology, then those two attributes will hardly seem like opposites. You'd probably think someone with a good memory for phone numbers would be *more* likely to notice small changes in appearance: she'd be detail-oriented, good at keeping track of small things. Certainly these don't seem like traits that would naturally be opposed to one another. But if you know something about the brain science behind autism, the fact that the two traits are inversely related makes perfect sense, because number skills and mindreading skills aren't simply the result of general intelligence; they're specialized modules, modules that for some as of yet unknown reason have been yoked together in the brain's wiring.

This is one of the key insights that neuroscience brings to our sense of self: strengths or weaknesses in one area are often predictive of strengths or weaknesses in seemingly unrelated areas. It makes intuitive sense to us that people who are better at processing language might be worse at processing visual data, or that blind people might have sharper hearing than people with eyesight. But you're less likely to get a nod of agreement when you propose that people who are good at factoring pi in their heads are usually bad at tracking eye movements. Yet that is the brain's reality. The more you understand the mind in the light of modern brain science, the more you recognize that isolated traits you possess aren't necessarily isolated—the brain is full of zero-sum games, where one talent prospers at the expense of another. Sometimes those balancing acts involve related skills; sometimes the connection is more obscure. Thus our final principle: *Your brain contains some strange bedfellows.*

Is mindreading one of our long-decay ideas, an idea that transforms your own sense of self? I believe it is, but to grasp that importance you can't think of mindreading simply as another word for "empathy." We all know people who are more empathic than others, who are more sensitive to others' feelings. Empathy is a powerful human trait, and it would be wrong to underestimate its centrality in our social interactions. By the same token, empathy is nothing new. What is new, I think, is the notion of the second-by-second, instinctive dance of mindreading: the mental sparring at the office party. Empathy is something you're consciously aware of feeling; you think to yourself, "It breaks my heart to see her so sad." Mindreading is faster than that, more invisible. The data it relies on flies by at lightning speed: a momentary tonal shift, a pause that suggests hesitation, a brief, inquisitive twist of the head. You may consciously evaluate the data once it has been interpreted—"Why did she seem startled by that news?"—but the act of interpretation itself is closer to a reflex than to a deliberate act of contemplation or analysis. One way of describing mindreading is via an idiom that we often use for performers: having a feel for your audience. Having a feel for your audience is different from being sensitive to the *feelings* of your audience, which is what empathy is all about.

For weeks after I first started reading about the neuroscience of mindreading, I found myself in conversations with friends or new acquaintances with a second-level, meta-interior monologue running through my head. Instead of watching their facial expressions for subtle clues about their internal state, I was watching their reactions to my expressions and speculating on their mindreading skills.

At a dinner party, I'd be listening to a friend follow a dozen irrelevant detours in telling what should have been a thirty-second story and suddenly recognize something I'd felt intuitively about him for years but never really put into words: he's mind-dyslexic. With other friends (many of them women) I finally understood part of why I had enjoyed our conversations so much over the years—our internal duets were as rich as the external ones. I put myself under the same microscope, noticing that in certain social situations I would be more "locked in" to my conversational partner, whereas in others my mindreading antenna appeared to get lousy reception. This resonance is the sign of a long-decay idea—it's like a tune that gets stuck in your head, and you can't help humming it wherever you go.

The more I thought about mindreading, the more I wanted to quantify my skills at it. The autism quotient test had whetted my appetite, but it was too subjective, and the skills it assessed were as much about that broader category of empathy as they were about the local reflex of analyzing facial expressions. I wanted my mindreading skills analyzed the way you'd have your vision tested, and I figured if there was anyone who could help me in this quest, it was Simon Baron-Cohen. That's how I eventually found myself scrolling through those computer images of eyes, scanning for drooping eyelids and furrowed brows.

I'd read a little about the eye-reading test before I actually sat down to take it and had imagined it to be much simpler than it turned out to be. Emotion scholars tend to divide up the spectrum of human emotion into two camps: the "primary" emotions of happiness, sadness, fear, anger, surprise, and disgust; and the "secondary" social emotions of embarrassment, jealousy, guilt, and pride. I figured the test would involve mapping one or the other of those ten sentiments to a pair of eyes, which seemed easy enough. But when I actually started to read through the instructions, I was shocked to find that the glossary of emotional states went on for several pages—ninety-three emotions in all, everything from "aghast" to "tentative." I'd anticipated choosing between "happy" and "sad," but instead the test wanted me to distinguish between "flirtatious," "playful," and "friendly," or "upset," "worried," and "unfriendly." As I read through the list, one disturbing thought came abruptly into my head: *I am going to flunk this test.* There was no way I could detect emotions this subtle in static images of two eyes. Perhaps my autism quotient score wasn't accurate, after all. If nonautistic people could read eye expressions at this level of sophistication, then maybe I was closer to Rain Man than I thought.

The test began with a grainy black-and-white image of an elderly man's eyes that looked like a close-up from a Jean Cocteau film. The left eye was wide open, the right more hooded. The emotion options were "hateful," "panicked," "arrogant," and "jealous." My first impulse was to choose "panicked," but as I studied that right eye, I began to have second thoughts. Was there something angry there? Or something wounded, as of a jealous husband who has just stumbled across his wife in the arms of another man? The more I scrutinized the image, the harder it got to discern a clear emotion. I decided to go with my initial hunch.

I turned to the next image, and a set of younger eyes of indeterminate gender stared back at me: perfectly symmetrical, with the slightest suggestion of a squint. I thought to myself, *This is what they mean when someone has a "gleam" in their eyes.* The first emotion option was "playful" and I immediately said, *That's the one.* But then I read on: "comforting," "irritated," and "bored" were the other options. Definitely not bored, but maybe what I saw as playfulness was really being comforting, being sympathetic. What was a gleam anyway? When I tried to locate the specific gleaming quality, the effect seemed to dissipate. As I searched for that original playfulness, I thought I detected a hint of irritation in the eyes. *This is*

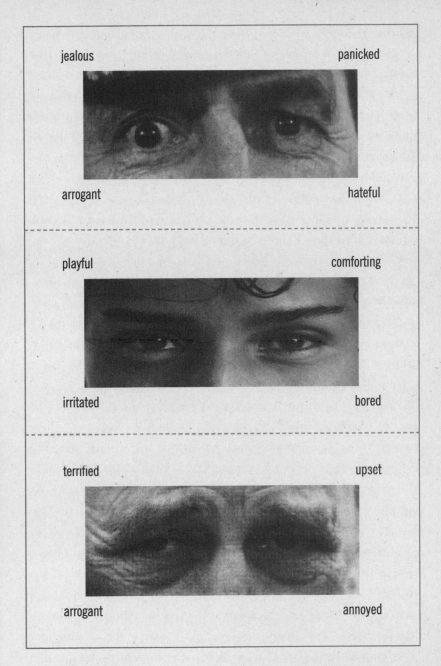

Three expressions from the "Reading the Mind in the Eyes" test.

madness, I thought: *I'm overanalyzing these images. Better to go with the gut, since this is supposed to measure gut responses anyway.* I marked down "playful" and moved on.

As the test progressed, I got a little better at sticking with my original hunches, but with each image, the clarity of the initial emotion grew less intense the longer I analyzed it. All but a few had an emotion that struck me at first glance, and while second thoughts caused me to doubt most of my first decisions, I went with my initial instincts throughout the test. By the end, I felt as though I would probably come out with half of the answers correct, which seemed like a pretty good ratio given the subtlety of emotions being presented.

But as it turned out, I was way off in my self-assessment. Instead of missing 18 of 36 questions, I had missed only 5. On the first seventeen images, the source of so much second-guessing, I'd been 100 percent right. It's an interesting test when you think you're failing, and you end up getting an A (or at least a solid B+). Particularly if you base all your answers on your gut reactions, and ignore all your attempts to outthink the exam. When I tried to interpret the images consciously, surveying each lid and crease for the semiotics of affect, the data became meaningless: folds of tissue, signifying nothing. But when I just let myself look—look without thinking—the underlying emotions came through with startling clarity. I couldn't explain what made a gleam *gleam,* but I knew one when I saw it.

If there was a connection to Rain Man's autism, it was here, in that instinctive "gut feeling," in the mental computation so fast and so transparent that it doesn't feel like thinking. Afterward, I was reminded of the classic stories of autistic people emptying a box of matchsticks and somehow just "seeing" exactly how many are scattered across the floor. The number just pops into their head, as vivid and unavoidable as a face. They have a gut feeling for numbers, the way most of us have a gut feeling for "playful" and "panicked."

Only neither feeling comes from the gut. After I finished the test, I asked Baron-Cohen what had been going on in my brain as I analyzed the images. "We've done fMRI scans of people taking the 'reading the eyes' test, and what we've found is that the amygdala lights up in trying to figure out people's thoughts and feelings. In people with autism, they show highly reduced amygdala activity," he explained. In many ways, the amygdala is the "gut feeling" center of the brain, implicated in all sorts of emotional processing. Recently, it has been shown to play a central role in our understanding of fear (which we will return to in the next chapter); when people have a "sinking feeling in their gut," or feel "gripped" by fear, the reaction has most likely been triggered by the amygdala. People with amygdala damage caused by strokes or head injuries often report that they are unable to detect fearful expressions in other people's faces. But as Baron-Cohen's test suggests, fear is only part of the amygdala story. "My hunch is that the amygdala is actually used to detect a much more varied range of emotions," he told me.

Inspired by the subtle emotional discrimination he found among his test subjects, Baron-Cohen has set out on a more ambitious quest: "We decided to figure out just how many emotions there are." He began with a survey of emotional descriptors taken from a collection of thesauri, which produced a list thousands of words long. Baron-Cohen and his team, aided by a lexicographer, then winnowed out the synonyms, creating a smaller collection of "discrete emotional concepts."

"We came up with a number," he said with a laugh. "Four hundred and twelve."

Four hundred and twelve unique emotions. The fact that our vocabularies include adjectives for so many emotional states, coupled with how well nonautistics score on the eye-reading exam, drives the point home: we are equipped biologically with an incredibly sensitive antennae for emotional variation. Baron-Cohen's latest mission is to build a tool that will help people whose antennae are broken. "What we've done is asked actors and actresses to create facial expressions for each of the four hundred and twelve emotions, and then included them all on a DVD. It's like an encyclopedia for emotion," he said.

"It was designed for people who score poorly on the autism tests, who want to learn emotional recognition in a slightly artificial way." Because autistics often possess higher-than-average skills at what Baron-Cohen calls "systematizing"—learning the rules of a given system, breaking it down into its component parts—one option for them is to improve their emotional recognition skills by systematizing the human face.

Baron-Cohen continued: "It's not the intuitive way of approaching people, but you could do it. You could try to figure out the rules that allow you to read another's emotional expression. It's like trying to learn a second language, sitting there with a grammar book and rules of syntax trying to figure it out in a different way than you would if you were a native speaker." The two approaches originate in different regions of the brain: the intuitive recognition centered in the amygdala and the systematizing ability residing in the neocortex, the seat of higher logic and language.

The clash between the amygdala and the neocortex explains my indecision while taking the eye-reading test. My gut reactions would flash up instantly from the amygdala, after which the neocortex would start analyzing the image in a more systematic way. But I haven't trained my neocortex to recognize emotions; I haven't spent time with Baron-Cohen's encyclopedia—precisely because my amygdala does such a good job on its own. And so the more I analyzed a given image logically, the less clear the answer became. The next time you're advised to trust your gut when you're meeting someone new, ignore the advice. Your gut has nothing to do with it. But by all means trust your amygdala.

There's a crucial scene near the beginning of Henry James's *The Golden Bowl* in which the recently married Maggie Verver walks in to find her beloved father, the long-widowed billionaire Adam Verver, engaged in what appears to be flirtatious conversation with a young woman. In a glance, Maggie suddenly grasps that her own marriage has created a new possibility: that her father might remarry after years of living as a bachelor with his only daughter. The rest of the book plays out, in a sense, the aftershock of this moment of recognition: the father does eventually marry another woman, with more or less disastrous consequences. But the originating scene itself unfolds without words spoken between father and daughter; it is as exacting, and as lyrical, an account of mind-reading as you are likely to find in literature:

> [Maggie's appearance] determined for Adam Verver, in the oddest way in the world, a new and sharp perception. It *was* really remarkable: this perception expanded, on the spot, as a flower, one of the strangest, might, at a breath, have suddenly opened. The breath, for that matter, was more than anything else, the look in his daughter's eyes—the look with which he *saw* her take in exactly what had occurred in her absence.

The visual communication flows in both directions; as Mr. Verver contemplates the look in his daughter's eyes, she in turn recognizes his recognition:

> He became aware himself, for that matter, during the minute Maggie stood there before speaking; and with the sense,

moreover, of what he saw her see, he the sense of what she saw
him. . . . Her face couldn't keep it from him; she had seen, on top
of everything, in her quick way, what they both were seeing.

James spends ten pages plumbing the depths of what he calls this
"mute communication"—slowing the tape down to analyze its
every twitch and unspoken innuendo. The passage gives us a wonderful instance of the human mind's powers of perception, on two
levels. First, there is the silent duet between father and daughter,
each of whom reads volumes into a simple pair of expressions
glimpsed across a room. And then we have the observatory power
of James himself, recognizing the depth of the exchange and drawing it out long enough for us to dissect its subtlety.

I bring up this scene because I think what James does here
runs parallel to what the brain sciences can do for our own self-
awareness. They can help us see our interactions with a new clarity,
to detect long-term patterns or split-second instincts that might
otherwise go unnoticed, sometimes because they operate below
conscious awareness and sometimes because we're so familiar with
them that they've become invisible to us. There are differences in
approach between the discerning eyes of scientists and novelists:
James doesn't offer a working theory to explain how Adam Verver
manages to gather so much information out of a passing glance;
and brain scientists don't usually weave their insights into gripping
narratives. But both approaches can illuminate the life of the mind.
To use a Jamesian term, they give us powers of discrimination.

In recent years, any time the brain sciences and the arts have
intersected, the debate has generally been framed in terms of evolutionary psychology: does the Darwinian approach have something useful to teach us about the cultural achievements of art? The
clashes that usually characterize these debates occur because on
some level evolutionary explanations operate against the grain of

art. Purely Darwinian models of the mind are about human universals, about what unites us as a species. Great novels or paintings or films are about the conflict between human universals and the local events of our personal and public histories. The narrative form that evolutionary psychology most closely resembles is myth: the enduring struggles and drives that define the human condition. The creative arts are about seeing what happens when individual lives intersect with these human drives, and often with the broader currents of history. This is why, more often than not, you get fireworks when the Darwinians and the art critics appear on the same panel. But when you widen the lens to see beyond evolutionary psychology, the conflict disappears. Brain science is as much about those chance events and individual personalities as it is about enduring truths and human universals. The last few decades of research have revealed, again and again, the way specific memories transform us as we grow and develop, the way life experience wires our brains as meticulously as our genes do. When we participate in mindreading's silent duet, we're drawing upon cognitive tools that are a part of our evolved human nature, but every mindreading exchange is also colored by the memories and associations unique to an individual life. We're wired to see smiles as a sign of internal happiness, but a smile can also remind us of a parent's grin from our childhood or a movie star's smile beaming down from the silver screen or a joke we told over breakfast this morning. Brain science has much to teach us about the way those individual memories are formed, and how they come to weigh on our subsequent behavior. The impact of past events on the present is so crucial to the modern understanding of the brain that this book doesn't include a single chapter on memory. This is because in many respects *all* the chapters are about memory.

Virginia Woolf described the compensation for growing old as gaining "the power of taking hold of experience, of turning it round, slowly, in the light." Memories transform our perception of the present, but the process is even more nuanced and layered than that: reactivating memories in a new context changes the trace of memory itself. For a long time, neuroscientists assumed that memories were like volumes stored in a library; when your brain remembered something, it was simply searching through the stacks and then reading aloud from whatever passage it discovered. But some scientists now believe that memories effectively get rewritten every time they're activated, thanks to a process called reconsolidation. (Freud sensed this process as well, though he gave it a different name: *Nachtraglichkeit,* or "retroactivity.") To create a synaptic connection between two neurons—the associative link at the heart of all neuronal learning—you need protein synthesis. Studies on rats suggest that if you block protein synthesis during the execution of learned behavior—the brain's memory of a reward cycle, for instance—the learned behavior disappears. Instead of just recalling a memory that had been forged days or months before, the brain forges the memory all over again, in a new associative context. In a sense, when we remember something, we create a *new* memory, one shaped by the changes that have happened to our brain since the memory last occurred to us. So the science is telling us two things: our brains are designed to capture the idiosyncrasies of our lives, and those lives—our memories of them—are being rewritten with each passing day.

You need only read a few pages of Proust to know that artists have been exploring these properties for centuries, if not millennia, just as James grasped the transformative power of mindreading. Indeed, the world of culture—particularly the poets and novelists and philosophers—has historically led the way in widening our understanding of the brain's faculties, much as that flower opened under Adam Verver's gaze. This they continue to do. The only difference now is that they have some competition.

2

The Sum of My Fears

A few years ago, my wife and I moved into an apartment in a renovated old warehouse on the far west edge of downtown Manhattan. The apartment was nice enough by New York standards, but it had one irresistible attraction: a massive eight-foot-high window looking out over the Hudson River. For the first few months in the space, any time interesting weather happened—a snowstorm, a particularly nice sunset—we'd gather together at the window and enjoy the view.

When summer arrived that year, we discovered a new feature of our view: thunderstorm watching. Most weather patterns blow in from the west where we live, and so as the temperature would start to climb, we'd see the thunderheads collect out over New Jersey, and we'd settle in for a good show. One mid-June afternoon, an especially severe storm started brewing—severe enough, we later found out, to cause the local news stations to run emergency bulletins forecasting heavy winds. As the skies darkened, and the whitecaps appeared on the river, we stood together at the window, faces practically pressed against the glass.

And then we heard a sound.

It was a subtle noise, like a twig snapping, and under the wind's whistling and the rumble of thunder it was hard to locate spatially. My wife shouted out: "What the hell was that?" She immediately jumped back from the glass, while I remained just to the side of the window.

In a typical display of composure and perceptiveness, I said I thought it might have been the door closing in the study, and my wife walked toward the back of the apartment to see if I was right. But that snapping sound wasn't, in fact, the study door; it was the bolt that anchored the bottom steel frame of the window. As I turned back to inspect the window, a sudden blast of air blew the frame right out of the wall, shattering a pane of glass the size of a kitchen table and sending shards through the entire length of the apartment. Because I was standing to the side of the window, the glass and frame blew right past me. If my wife had not ventured off to inspect the study door, she would have taken the full force of a steel frame and glass panel being blown in by sixty-mile-an-hour winds. It's entirely possible that the impact would have killed her.

You may know what those few seconds felt like. First, there is the extraordinary sense of time slowing down. The window itself probably took less than a tenth of a second to blow out, but I have a distinct memory of thinking—all before the glass hit the floor— that physically we were going to be all right, that my wife was too far away to be injured. A split second later, I was wondering if perhaps the storm had kicked up a tornado, which meant that standing next to an open window a hundred feet aboveground wasn't exactly the safest place to be. Within a few seconds, we'd barricaded ourselves in a bathroom, and only then did I notice that my heart rate had increased and my palms were sweating. Below my direct awareness, blood had rushed to my extremities, preparing me for sudden movement and creating the "nervous stomach" feeling in my gut. My adrenal glands had secreted a sudden rush of adrenaline, which helps prepare the body for sudden movement by converting glycogen into energy-rich glucose. My reflexes had been primed, making me much more likely to startle in response to another unexpected sound; pain sensations were dampened.

But what I noticed most at the time was an otherworldly sense of alert clarity. I remember thinking: *If Starbucks ever figures out a way to produce this feeling via a cup of espresso, they'll really take over the world.*

This is the body's fear response, an orchestral mix of physiological instruments launching with masterful speed and precision. We talk about it colloquially as the fight-or-flight response. Feeling it kick in is one of the best ways to experience your brain and body as an autonomous system, operating independently of your conscious will. A second after the window blew in, I made a deliberate decision to hide in the bathroom for protection, but my brain was making comparable decisions as well—decisions executed in a purely unconscious, instinctive fashion but nonetheless designed to protect me from harm. The effects of those decisions are as powerful as those of many recreational drugs, though they are concocted entirely by the brain's internal chemistry.

The fight-or-flight response is amazing, but it is also old news. Even before we started monitoring our fluctuating serotonin levels or training the right side of our brain, the phrase "adrenaline rush" was part of the popular vocabulary. The immediate physiological reaction of fear is familiar enough. Fight-or-flight launches in a matter of seconds, and its effects can dissolve in minutes. But the trace memory of that fear can last a lifetime.

* * *

I have a phobia of bees. At an earlier point in my life, it was a genuine social impediment. (Picnics were off-limits, and during yellow jacket season I was usually huddled indoors while my peers enjoyed the brilliant New England fall weather.) I also have a mild fear of heights and a fear of flying that cycles on and off every few years, usually in sync with world events. Chances are, you have an equivalent cabinet of horrors yourself. Some might have been triggered by events in your personal history, while others may be traced to genetic origins. I suspect my fear of heights is more biology than personal history, but I know that my fear of bees dates back to a series of traumatic stings that I experienced as a child. With the airplanes— Well, if you read the papers or watch the news, you can't help being a little afraid of strapping yourself into a flying tube loaded with explosive fuel.

But ever since that June storm, a new fear has entered the mix for me: the sound of wind whistling through a window. I know now that our window blew in because it had been installed improperly, with a single joint to bear a load that was supposed to be distributed between two—that snapping sound we heard was the one joint giving way. I am entirely convinced that the window we have now is installed correctly, and I trust our superintendent when he says that it is designed to withstand hurricane-force winds. In the five years since that June, we have weathered dozens of storms that produced gusts comparable to the one that blew it in, and the window has performed flawlessly.

I know all these facts—and yet when the wind kicks up, and I hear that whistling sound, I can feel my adrenaline levels rise. If I'm sitting beneath the window, I have to move to the other side of the room before I can concentrate on anything other than the whistling. And even positioned safely out of harm's way, I have an edgy, background sense of dread until the wind dies down again. Part of my brain—the part that feels most *me*-like, the part that has opinions about the world and decides to act on those opinions in a rational way—knows that the windows are safe, and short of a Level 3 hurricane, I should be able to look out at the Hudson with perfect ease. But another part of my brain wants to barricade myself in the bathroom all over again.

Fear memories like this one are another way of apprehending the brain's modular nature directly. Even mild trace memories of traumatic events can leave you feeling a little like a split personality, while serious cases of post-traumatic stress disorder can be horribly debilitating, particularly in the face of stimuli that somehow resemble the original trauma. The flight-or-flight response may offer a vivid example of the body's physiological instincts, but in the classic examples—as the mugger chases you down the alleyway or the bombs scream across the sky—your instinctive responses are in sync with your rational ones. Part of your brain says, "I'm afraid," and the other part says, "For good reason." But years later, when your heart starts racing at the sound of whistling wind, a gap opens up between your conscious assessment of the present danger and some other assessment elsewhere in your mind. Part of you knows that you're safe, and part of you can't help being afraid. Which you is you?

Answering that question takes us back nearly a hundred years to a French psychologist named Edouard Claparede, who was treating a woman suffering from a rare form of amnesia that left her incapable of forming new memories. Claparede's patient had suffered localized brain damage that preserved her basic mechanical and reasoning skills, along with most of her older memories. But beyond the duration of a few minutes, the recent past was lost to her—a condition captured brilliantly in the recent thriller *Memento*, in which a man suffering from a similar ailment attempts to solve a mystery by furiously scrawling new information on the backs of Polaroids before his memories fade to black.

Claparede's patient seemed herself straight out of some kind of slapstick farce, if only her condition weren't so tragic. Each day the doctor would greet her and run through a series of introductions. If he then left for fifteen minutes, he would return entirely unrecognized. Doctor and patient would do the introductions again, and the patient would find herself conversing with a brand-new doctor. One day, Claparede decided to vary the routine. He introduced himself as usual, but concealed a thumbtack in his palm the first time he reached to shake her hand.

It doesn't sound like your ideal initial consultation with a physician, but Claparede was onto something. When he arrived the next day, his amnesiac patient greeted him with the usual blank welcome—no memory of yesterday's pinprick, no memory of yesterday at all. Until Claparede extended his hand. Without being able to explain why, Claparede's patient refused to shake hands with her doctor. The woman incapable of forming new memories had nevertheless remembered something from the recent past—a borderline-conscious sense of danger, the trace memory of past trauma. She failed utterly to recognize the face and the voice she'd encountered every day for months. But she remembered the fear.

Why would an amnesiac suddenly develop a flair for memory over a thumbtack? Until recently, Claparede's patient seemed to scientists like a complete anomaly, one of those rogue data points that made no sense under reigning scientific orthodoxy. For most of the last fifty years the story was the same: the brain relied on a general intelligence, increasingly imagined as a computer, that drew upon past experience to make rational assessments of new situations. This process was the basis of learning and emotion. You had inputs from the outside world; you had memories of past inputs; and you had some kind of glorified calculator that would measure these inputs against each other and come up with a behavioral strategy. If you experienced fear in the face of a given stimulus, it was because your memory bank had pulled up some past experience of danger that resembled the present stimulus. The emotion of fear itself was a secondary effect, like a command issued from the rational brain: "The data suggest that there's cause for fear here, so we should start feeling afraid now."

Claparede's patient threw a giant monkey wrench into this model. How could someone learn fear when her memory system couldn't learn at all? It would be like me developing a phobia of wind without remembering our window crashing in. You can't build a rational assessment of potential danger if your memory banks aren't supplying any information about past encounters. If fear resulted from a rational assessment of risk, how could Claparede's patient both know to avoid her doctor's handshake and yet not know why? Clearly, her brain had captured some trace of the pinprick in some alternative storage system unavailable to consciousness. But where was the memory stored?

Learning to be afraid turns out to have been one of the most studied behavioral patterns of the twentieth century. In fact, Claparede's thumbtack is itself a somewhat fiendish twist on a classic behaviorist experiment, one almost as famous as Pavlov's canine dinner bell: fear conditioning. Put a rat in a cage, play a tone, and simultaneously deliver a shock to the animal. After even a single pairing of tone and shock, the rat starts to fear the tone itself, even if it's not always accompanied by the shock. The fear reaction is known as a conditioned response: the rat has an unconditioned, innate fear of shocks, and it can be conditioned to be afraid of tones if the two are associated with each other. Claparede's thumbtack is the shock. His outstretched hand is the tone. After only one exposure, the amnesiac patient acquired a conditioned fear response to shaking hands with her doctor.

As their name implied, the behaviorists were interested in the hard data of external appearance: whether the rat acted afraid or not after a few sessions of tone-and-shock. At heart, they believed that you *could* judge a book by its cover. How the rat's brain actually processed that fear response and how that fear response felt subjectively to the rat—these questions were off-limits to the behaviorists, sequestered away in the "black box" of the unknowable brain.

But over the past few decades, science has opened up that black box, and charted the actual pathways of conditioned fear in the brain. The leading figure in the field is a soft-spoken, if controversial, NYU professor named Joseph LeDoux. When I visited him at his office near Washington Square, he told me the story of his intellectual quest. It has the classic rogue-science beginning. "My first grant request on this topic in the early 1980s," he explained with a smile, "was turned down because emotions can't be studied scientifically."

In its initial phase, LeDoux's research was cartographic in nature, a literal mapping of the mind. But because advanced brain-imaging technologies weren't yet available, LeDoux's mapmaking was more Lewis and Clark than GPS satellite. The tone-and-shock fear-conditioning experiment gave him a simple causal chain to explore, because the pathways of sound-processing in the rodent brain had already been delineated. He knew where the sound entered the brain and where it was integrated into conscious perception, and he knew several key relay points along the way, including three primary stops: the initial data processing performed in the brain stem, followed by one of the brain's primary hubs, the thalamus; only after passing through those initial two stopovers did the sound enter conscious awareness, in the auditory areas of the cortex.

LeDoux's approach was a kind of surgical subtraction. Take a healthy rat and begin extracting specific parts of its brain; if you remove a region and it can still learn to associate the tone with the shock, then the region you've removed isn't necessary for fear conditioning. But if the rat stops learning, you know you've got something relevant.

LeDoux began at the end of the chain: the auditory cortex, where the sound, having traveled through various way stations starting with the eardrum, is finally integrated into our sensory experience of the world. When he removed that region, the rats could still learn to fear the tone. With the auditory cortex removed, they were an even more extreme version of Claperede's patient: afraid of a noise without even hearing it. So the learning didn't reside at the conscious end of the chain; it was somewhere in the middle. "So I went down one station to the auditory thalamus," LeDoux said. "I took that out, and they couldn't learn at all. So that meant that the sound had to go through the system to the level of the thalamus, but didn't go through the cortex. So where was it going?" The question was a puzzling one, because the auditory thalamus was supposed to be just a relay station from the ear to the primary destination, the auditory cortex. That implied something strangely inverted about LeDoux's result: you could eliminate the primary destination altogether without affecting the learning, but if you took out the relay station, the learning stopped.

LeDoux's assumption was that the auditory thalamus harbored a link to another part of the brain, in addition to its link to the cortex. Using a tracer dye to follow pathways out from the auditory thalamus, LeDoux discovered a connection to the amygdala, the almond-shaped region in the forebrain that we encountered in the "Reading the Mind in the Eyes" test. When he removed the amygdala from the brains of the rats, they failed to learn. Subsequent experiments also demonstrated that a crucial part of the amygdala known as the "central nucleus" contained links to key brain stem areas that controlled the autonomic functions involved in fight-or-flight, like heart rate acceleration. "I didn't start out looking for the amygdala," LeDoux said. "The research led me to it."

The principal insight that emerged out of LeDoux's research is that the experience of danger follows two pathways in the brain, one conscious and rational, the other unconscious and innate. These were quickly dubbed the "high road" and the "low road." Say you're walking through a forest and out of the corner of your eye you detect a slithering shape to your left, accompanied by a rattling sound. Before you even have time to formulate the word "snake," your body has frozen in its tracks; your heart rate has accelerated; the sweat glands on your palms have dilated. In your brain, the information flow looks something like this: your eyes and ears transmit basic sensory information to the auditory and visual thalamus, where the information is then transmitted along two paths. The first path heads toward the cortex, where it will be integrated with other real-time sensory data, along with more elaborate associations: the word "rattlesnake" itself, or your childhood memories of a pet python, or the snake scene from *Raiders of the Lost Ark*. Around the same time, the slithering rustle would also be transmitted—in less rich detail—to the amygdala itself, which would blast out an alarm to the brain stem, alerting the body that a potential threat is nearby. The key difference between the two paths is time: via the high road, it might take a few seconds to establish the presence of the snake and formulate a response, while the low road kicks the body into a freezing response within a fraction of a second. Of course, none of the elaborate bodily choreography involved in this response has to be learned, the way you might learn a complicated yoga position; your body knows how to execute the freezing response without any training at all. In fact, it knows the response so well that it is nearly impossible to keep it from happening in the face of a sudden threat.

The memories captured by the amygdala during traumatic events have two intriguing properties. The first is that they contain less information than traditional memories—what memory researchers call "declarative memories." The visual cortex perceives the snake in all its slithering glory; your declarative memory of the forest snake might include a distinct pattern on its skin or the exact series of movements it made before you turned and ran. But your amygdala retains a much cruder portrait, as though the event were being shot by a Polaroid instead of an IMAX camera. The amygdala might store only the general slithering movement and the long, thin black outline of the snake's body against the grass. A higher-resolution image would exceed the capacity of the channel that connects the thalamus to the amygdala, so what you get is a quick but dirty image rather than the slower, but more realistic, portrait created by the visual cortex. (The same goes for information transmitted by the other senses as well.) This "quick sketch" approach helps the body respond to threats with extraordinary speed, but it has a troubling side effect. Because fear memory is fuzzier than declarative memory, the pool of objects that potentially resembles the specific fear memory is much larger: a dark stick or a garden hose in the grass can easily fool your amygdala into thinking that you've stumbled across another rattlesnake, even if your visual cortex can easily tell the difference.

The quick sketch approach explains partly why post-traumatic stress disorder is such a difficult condition to treat. The amygdala of a Vietnam vet may hear an AK-47 every time a truck backfires, and every thunderstorm sounds like carpet bombing. If the amygdala could somehow be trained to discriminate better, these flashbacks wouldn't happen as frequently.

Yet a virtue lurks in that lack of discrimination. If your trauma memories were too specific, your brain would be incapable of learning from experience—or, more precisely, incapable of learning general principles from experience. If a more discriminating amygdala encountered a rattlesnake with brown spots on its skin, it might not know to be afraid of one without spots. Snakes that approached you from the left side might leave you unafraid of snakes approaching from the right. The quick sketch approach to fear memory lets you move beyond the details into general rules of thumb: if you see something slithering in the grass—spots or no spots—then run away. In Borges's classic story "Funes the Memorious," the protagonist is gifted (and cursed) with an uncanny, beyond-photographic memory capable of conjuring up the slightest details from the most incidental occurrence two decades before. Near the end of the story, the narrator says, "I suspect, nevertheless, that [Funes] was not very capable of thought. To think is to forget a difference, to generalize, to abstract. In the overly replete world of Funes there were nothing but details . . ." In reducing the resolution of the fear memory, the amygdala performs a kind of thinking, searching out underlying commonalities in a world of unique threats.

The other intriguing property of fear-learning is what some brain scientists call "flashbulb memory." During traumatic events, your brain stores not only a trace of the specific threat—the snake, or the oncoming car, or the rattle of AK-47 fire—but also the contextual details surrounding that threat. This is a classic expression of the brain's associative architecture, captured by the famous slogan "cells that fire together, wire together." Different incoming stimuli trigger activity in specific constellations of neurons; when those neurons fire in sync with one another, they are more likely to form new connections. As the connections grow stronger, a given neuron has an easier time triggering another connected neuron. This process—the root of all synaptic learning—goes by the name "Hebbian learning," after the Canadian psychologist Donald Hebb, who first proposed the model in 1949.

Think of a traumatic event from your own past, one that involves a sudden danger, like a car accident. You no doubt remember the immediate threat—headlights bearing down on you or a screeching tire—but most likely you also possess a number of extraneous memories: the song playing on the car stereo at the moment of impact, the color of the early-evening sky, the confused expression on the face of an onlooking pedestrian. None of those details actually seems related functionally to the threat posed by two cars colliding with each other, and yet when you hear the song five months later, you can feel the fear response welling up inside you. The neurons associated with the screeching tires had fired at the same time as those associated with the song on the radio. The fear response helped wire them together.

Once again, a lack of discrimination has a potentially adaptive value. In life-or-death situations, you never know where relevant information might lie. Say that your rattlesnake encounter takes place next to a forest stream. The sound of a brook babbling isn't relevant to the rattlesnake threat, but a rattling sound is emphatically so. Our brains are designed to take note during traumatic events of all sensory inputs—albeit in a low-resolution form—on the off chance that some stray element will turn out to be a good predictor of future threats. If that means we develop irrational fear of babbling brooks, and an entirely rational fear of rattling, so be it. The irrational fear won't kill us, but not acquiring the rational fear very well might.

This, too, is a kind of thinking. In the months after 9/11, I started noticing a subtle but predictable shift in my general levels of Manhattan-dwelling anxiety. Crisp, clear weather made me more nervous than overcast days. For a long time, I thought this was purely extraneous associative learning: September 11 itself had been a spectacularly clear day, which is one reason my memories of standing on Greenwich Street and watching the towers burn was so vivid—there was no moisture or smog in the air to block the view. So when similar weather elevated my anxiety levels, I thought of it as being like that song on the radio during the car accident: a stray detail, unrelated to the real threat, that nevertheless becomes associated with the fear memory.

But then one day, while walking along the same path I had followed on the morning of the attacks, I had a small epiphany. I realized that my amygdala had stumbled across a clue that hadn't occurred to my rational brain. Forget about all the other threats that arose in the public imagination after 9/11—anthrax and dirty bombs and smallpox—and think exclusively of the specific assaults that took place on that horrible day. If the threat that your brain is trying to protect you from is hijacked airplanes flying into skyscrapers by visual flight rules, then cloudy days *are* probably less dangerous than clear ones. If it's hard enough to hit a building without a flight plan in perfect weather, then it's almost impossible to do it when half the building is concealed by fog. If the imminent danger was an exact replay of the 9/11 attacks, there was nothing *irrational* about being more anxious on sunny days.

I made a number of conscious evaluations of potential threats after 9/11—I avoided densely populated parts of the city and tall buildings whenever possible. I drove or took the train when traveling along the Eastern Seaboard, forswearing the shuttle flights that had been such a staple of my life over the previous decade. These were deliberate strategies for dealing with a possible attack, developed by analyzing the patterns of a past one. But my amygdala was also evaluating the danger, and creating its own strategies. And one of those strategies was: be on the lookout during nice weather. Of course, the amygdala wasn't actually working through the logic on its own; it simply stored a flashbulb memory of that day, and one of the elements illuminated was the brilliant blue sky. When subsequently my amygdala encountered similar skies, it set off an alarm. I had missed the connection between the weather and the attacks in my subjective appraisal of that nightmare day. But my amygdala hadn't.

* * *

Where are these quick-and-dirty flashbulb memories stored? Some scientists believe the amygdala doesn't have its own discrete storage system for emotionally charged memories but rather somehow marks memories created by other parts of the brain as being emotionally significant. In 2001 James McGaugh of the University of California at Irvine conducted a telling variation on the classic fear-conditioning experiment by delivering a shock to a rat if the animal took a step. After administering the shock, McGaugh injected cyclic AMP—a cellular messenger that strengthens neuronal synapses, leading to stronger memory—into the animal's cortex. Two days later, the rats were tested to see how well they had been conditioned; those that received the cortex injections turned out to have enhanced memories of the shock. "So we know the cortex is involved in the memory that's based on fear in that situation," McGaugh tells me after I call him up to discuss the experiment and its implications. "Now, if we make a lesion of the amygdala, the stimulation of the cortex doesn't do anything. In other words, you have to have a working amygdala for the cortex to do its job."

I ask him about the implications of these results. "That experiment tells me that fear is not learned *in* the amygdala," he explains. "Amygdala projections are coming up to brain regions where information is being stored, and they're saying: 'You know this memory you're storing? Well, it turns out to be a very important one, so make it a little stronger, please.' It provides selectivity in our lives. You don't need to know where you parked the car three weeks ago, unless it was broken into that day." You can think of this selectivity as the brain's way of underlining.

At first glance, this might seem like another instance of the neuromap fallacy. Why should it matter to you whether the memory is stored in the amygdala or not?

Two reasons. First, if the memory is stored in some secure, undisclosed location of the mind, inaccessible to conscious awareness, then all sorts of possibilities for psychological dysfunction open up, because the memory has a dual life in your brain. The cortex can forget, but the amygdala can keep the fear alive, albeit somewhere below the radar of awareness. Before long, you're finding yourself afraid of all sunny days everywhere, and you have no idea where the phobia originated.

More important, though, if the amygdala is merely underlining important memories stored elsewhere, then the brain science is potentially telling us something new about the way we handle traumatic memories, particularly after an episode itself has passed. It's the amygdala activation that underscores the memories, and from the memory's point of view, it doesn't really matter whether the underlining is being triggered by the actual event or its recollection. If your body loads up with the fight-or-flight response, the memory grows more pronounced—even if you're simply *recalling* events that took place in the past.

"Say you have a traumatic experience," McGaugh says. "The memory of that experience will pop into your brain the next day, whether you want it to or not. And when that memory pops into your brain, you're going to have that whole autonomic response that you had originally. It's going to come back again. So it's not only that you remember that you were mugged, but you also get very emotionally excited about it when the memory happens." That emotional excitement triggers the memory-enhancing cycle all over again, making the traumatic memory even stronger, like a spinning tire deepening the muck hole it's stuck in with each jab to the accelerator.

As McGaugh is telling me about this, I think back to the days and weeks that followed our window blowing in, during which time I found myself replaying the event again and again in my head, with one horrifying alteration: my wife doesn't jump back from the window at the sound of the joint snapping. Instead she stays at the glass for five seconds longer, and the window crashes in with her standing beneath it. Just contemplating the thought for a split second would fill me with gut-tightening dread, but I couldn't help returning to it. The chain of events seemed far too fragile: take away that one brief, instinctive decision to move away from the glass, and our lives are potentially altered forever. There was too thin a line separating what happened from what might have happened, and each time I contemplated that line, I could feel my body's stress response relive the event all over again.

This gets to the heart of why McGaugh's research is not just another case of gratuitous neuromapping. Knowing about the way fear memories are stored suggested something genuinely new to me about the way my wind phobia came into being. The fear began, of course, with the initial event, but the phobia itself may well have been cemented by my subsequent ruminations on the disaster that might have happened. If I had managed to avoid replaying those bleak scenarios in my head—or at least kept my amygdala from triggering the fear response when I replayed them—the phobia might not have developed at all. I would still remember the event, of course, but the sound of air whistling through a window wouldn't start my heart racing all over again. If after the fact you can keep your amygdala from underlining the memory, you can potentially ward off a phobic reaction.

One way of preventing the underlining is through drugs. Beta-blockers prevent the body's autonomic system from kicking in during stressful events. (People who suffer from a fear of public speaking sometimes take these drugs before giving a speech to keep their heart rate down.) In a few recent studies inspired by McGaugh's findings, recent trauma sufferers have been given beta-blockers. By preventing the autonomic reaction, beta-blockers keep the memory from forming deeper grooves in the brain, making post-traumatic stress symptoms less severe.

All of which raises the question of whether a little old-fashioned repression might be a healthy response for people in the weeks after experiencing a traumatic event. It's a popular cliché that we need to "work through" frightening experiences, through therapy or extended conversation with loved ones. But if replaying the event, and consequently triggering the body's autonomic response, can lead to later stress disorders, then maybe the worst thing you can do in the immediate aftermath of fear is talk through the incident. Maybe you're better putting it out of your mind altogether, at least until the fear response subsides.

The more I learned about the amygdala, the more it seemed to me that this small region of the brain deserved to be a household term, recognized as widely as the "natural high" endorphins or serotonin. If the opposition between left and right brain had become the subject matter for self-help books on management strategies or learning how to paint, then the amygdala deserved an even more prominent place in the spotlight. When you're intuiting a spouse's bad mood from a subtle look in her eye, or you're recovering from a terrifying skiing accident, or you're battling your phobia of snakes—in all these cases, your amygdala directs your appraisal of the world, whether you're aware of it or not. The more I learned, the more it seemed to me that a fundamental tension in the human brain lay in the battle between amygdala and neocortex—the emotional center wrestling for control of the organism with the seat of reason. The trouble with emotional memories is that they can be fiendishly difficult to eradicate. The brain seems to be wired to prevent the deliberate overriding of fear responses. Although extensive neural pathways link the amygdala to the neocortex, the paths running in the reverse direction are sparse. Our brains seem to have been designed to allow the fear system to take control in threatening situations while preventing the reign of our conscious, deliberative selves.

This design may have been optimal for predator-rich environments in which survival was a minute-by-minute question, but it is not always a useful adaptation for modern environments in which the stressors can be job-performance reviews. The amygdala may be looking out for your best interests by preserving a memory of a window blowing in, but if the end result is an inability to sit still in your apartment during 20-mile-per-hour gusts, the fear circuitry has gone too far.

It's tempting to see the battles between these different regions as a reenactment of Freud's proposed clash between man's civilized superego and his primal id. Certainly knee-jerk antipathy to the Freudian unconscious—an unwillingness to accept the premise that we are shaped by drives outside of our conscious awareness, drives that often run counter to our perceived interests—is dead wrong. Our brains teem with crucial processes that run below the radar of our surface perception—and it's a good thing, too! If we had to deliberately parse out other people's emotional nuances, or evaluate new situations for potential threats, we'd never get anything done. We're better off letting the amygdala do that work for us, even if we're usually not aware that it's working at all.

So in our modern understanding of the amygdala, there is some vindication for the Freudian model. But in other crucial respects, the connection seems less fitting. Think about the watered-down Freudian language that we now use to discuss trauma and memory. We're all aware that traumatic events play tricks on our normal memory, and that certain elements of those memories hang around below our awareness for long periods of time, surging up at bizarre moments. We also recognize that our minds can sometimes return insistently to traumatic events, usually against our will. But our explanations for these phenomena can be peculiar. When some stray detail from a long-forgotten trauma pops into our head, we assume that somehow the original memory in all its clarity has been *repressed* by the psyche, beaten down into submission. The phobic details that reappear years later enact the return of that repressed memory. You don't remember the snakebite from childhood because it left too strong an impression for your brain to handle, but when you flinch at the sight of the garden hose, that repressed memory is clawing its way back into consciousness.

But in the modern portrait of amygdala activation, there is no clear mechanism for the repression or censoring of traumatic events; there are unconscious memories recorded, but those memories are not unconscious from having been repressed by some kind of internal censor. They're unconscious because the amygdala operates largely below conscious awareness, and regulates autonomic behavior that we can't directly control. The traumatic memories are captured by the amygdala not because the executive brain (the ego, in Freud's formulation) can't tolerate them in some fashion; they're captured because they contain information that might be useful to the future safety of the organism. It's good for the organism's future prospects to have a quick-and-dirty fear response to the sound of a rattlesnake, a fear response that doesn't need the slower pathways of conscious processing. The cause of memory storage by the amygdala is not some kind of internal censorship; it's efficiency. Some normal declarative memories, however traumatic they might be, just fade away over time, but the amygdala is more tenacious. It keeps score. If a snake bit you in your early teens, the amygdala may well retain a crude trace memory of that event for decades after the traditional memory has faded. In a sense, we are all a little bit like Claparede's amnesiac patient: remembering the fear more vividly than the original trauma.

There is some evidence, largely derived from the study of rats, that severe stress may impede the formation of declarative memories. Extended release of the stress hormone glucocortocoid causes atrophy in the neurons of the hippocampus, though the effect is reversible if the stress comes to an end. Long-term stress may cause permanent damage to the hippocampus. So there may well be a neurological explanation for why traumatic memories have a tendency to disappear from consciousness while persisting in our "gut" reactions and phobias. The stress response weakens your hippocampus so much that the declarative memory never forms, while the amygdala manages to capture the traumatic event via the "low road." You have an emotional memory, but not a declarative one. But while the trauma registers only in your unconscious perception of the world, the declarative memory is not the victim of repression, in the strict Freudian sense. It is closer to the temporary amnesia you experience after you've taken a blow to the head. Activation of the body's stress system—like a blow to the head—has all sorts of debilitating physiological effects: elevated blood pressure, higher rates of heart disease and even cancer. One of those effects is impaired memory. When a baseball player is hit in the temple by a fastball and wakes up without a memory of the pitch, we don't say that the memory was too traumatic to be processed by the mind's bureau of standards and thus had to be repressed. We accept that a projectile traveling at 90 miles per hour will damage the neurons dedicated to recording memories. Where lost memories are concerned, severe stress is more like that fastball than an internal censor.

What about traumatic memories that we find ourselves revisiting compulsively? Here our pop psychological assumptions follow a slightly different path, one that dates back to the model Freud developed in *Beyond the Pleasure Principle* after treating a number of patients struggling with traumatic memories of carnage on the battlefields of WWI. In the book, he famously amended his theory of the human mind's underlying drive for pleasure with an additional, sometimes contradictory "death drive," in which the organism seeks out, above all else, the cessation of stimuli. Freud had been forced into this reformulation by the seemingly illogical mental behavior he kept witnessing among the veterans: an often debilitating insistence on revisiting the traumatic memories of war in flashbacks, dreams, panic attacks set off by loud noises, and so on. (The term "post-traumatic stress disorder" didn't exist at the time, but that's exactly what Freud was documenting.) This repetition-compulsion made no sense in a psyche driven by the pleasure principle alone, particularly in dreams, which were supposed to be nonstop pageants of wish fulfillment. And yet the trauma sufferers' dreams returned endlessly to the frontlines, replaying in horrifying detail the brutality of war. If the mind simply wanted to feel pleasure all the time, why would it bother reliving these terrible memories?

Part of Freud's solution was to suggest that reliving these memories was the psyche's way of subduing them; by recalling such horror at will, the mind somehow brought the events under conscious control, thus making them less threatening. "These dreams are endeavoring to master the stimulus retrospectively," he wrote. Predictably enough, Freud saw a similar pattern at work in a young child's anxiety about being separated from his mother. He tells the story of watching his grandson playing a game of "fort-da" incessantly: hiding a toy, and then uncovering it, then hiding it again, then recovering. By repeating the trauma of loss and subsequent recovery, the boy was mastering the experience and diminishing the psychic trauma it inflicted on him.

But Freud also went beyond the idea of simple mastery. The veterans weren't just revisiting wartime events to gain control of the memories, to subdue them; they were also venturing back to those days because they possessed a crucial drive as part of what Freud called the "psychic economy," an underlying desire to restore the self to that original state of total quiescence. Absence of all stimuli was the goal, Freud argued, not of positive or negative stimuli. This was the death drive operating alongside the pleasure principle.

Freud's "death drive" solution to the repetition-compulsion problem had an enormous impact on both scientific and popular perceptions of how the mind works at times of extreme stress. But his theoretical gymnastics seem superfluous in the light of modern brain science. The brain is not trying to release what would otherwise be overwhelming energies—mastering the trauma by letting off steam—when it revisits traumatic events against the conscious will of the individual. The brain is revisiting these memories because, over the millions of years that our central nervous system has been evolving, a mental circuit was created that (1) recorded details from traumatic encounters and (2) set off a systemwide alarm when it was reminded of those details. That circuitry helped our ancestors survive and pass on their genes more often than not, which is why we find the amygdala and its fear responses in so many species. Freud was exactly right to insist that certain regions of the brain operated outside of our conscious control. But there is nothing in the brain's architecture, as we now understand it, that compels the organism toward death. (There may be a death drive in our *genetic* architecture, in the form of body clocks that tell our organs when to stop doing regular maintenance work. But that's another story.) The fear response—however incapacitating it can be—is fundamentally about staying alive. It's about staying alive when there isn't time to think.

The brain evolved a strategy to cut conscious thinking and decision-making out of the picture and to let the amygdala do the work. When the brain revisits traumatic memories, it isn't trying to beat them into submission; it's trying to keep them relevant. You're not reminded of frightening events because the brain is somehow drawn to death imagery or because it wants to eliminate the memory by replaying it ad nauseam. The memories come back to you because, on some basic level, they're good for you. It seems unnecessary to be afraid of things that end up being harmless 99.99 percent of the time. But the 99.99 percent of unwarranted fear is merely a passing annoyance compared to the threat posed by the remaining .01 percent. I don't like being afraid of wind, but being afraid is not going to kill me. Flying windows, on the other hand, just might.

3

Your Attention, Please

I am pedaling a bicycle with my brain. Or, more accurately, I am failing miserably at pedaling a bicycle with my brain.

I'm high above Central Park on a sultry August afternoon, sitting in a suite at the Mayflower Hotel, where a team of executives with a company called the Attention Builders have been holding an all-day training session for use of their new product, which happens to be strapped to my head. The product—part of an integrated system called the Attention Trainer—looks like your standard Day-Glo bike helmet, but it harbors state-of-the-art neurofeedback technology that measures changes in the electrical activity of certain sections of your brain and reports them back via a wireless connection to an ordinary PC.

The Attention Builders, as the name suggests, have built this system to help children combat that chronic end-of-millennium complaint, attention deficit disorder. The helmet tracks specific types of electrical activity associated with ADD (and the related disorder ADHD, short for attention deficit hyperactivity disorder). The data generated by the helmet is presented graphically in real time to the person wearing the helmet. Because the company is, literally, trying to capture the attention of kids with this product, the Builders have concocted a series of video games that respond to the data generated by your brain, games that reward high-attention states and discourage more distracted ones. Start zoning out while connected to the Attention Trainer software, and you'll see it reflected on the screen within a split second. Start paying attention, and you'll find yourself winning the game.

And that's my problem. I'm losing the game.

Neurofeedback dates to the late 1960s and early '70s, during which time it experienced an initial blip of Aquarian hype, alongside primal scream therapy, TM, and EST. Early advocates found that the technology enabled users to reach meditative states more easily by encouraging the alpha brain waves associated with deep relaxation. The trouble with the first neurofeedback machines revolved around the weakness of the feedback: lacking modern graphical displays and high-speed processors, the early units represented brain activity with squiggly lines and R2D2-like computer bleeps. While the evangelists talked about using neurofeedback to fine-tune the instrument of your mind—the seminal history of the movement is a book called *A Symphony in the Brain*—the technological limitations of the age made that instrument sound about as sophisticated as a cell phone ringer.

But it took more than an increase in computational power to usher in a neurofeedback revival. It's one of many small ironies in the history of neurofeedback that a technology originally associated with the alternative mind-expansion movements of the '60s should find new life thanks to today's culture of grade-school overachievement. While neurofeedback machines are being used in a variety of contexts, the most common use today is therapeutic, for treating ADD and ADHD as a nonchemical alternative to Ritalin and other prescription stimulants.

"I had gotten interested in neurofeedback in the late nineties," Attention Builder CEO Tom Blue tells me at the Mayflower before I put on his device. "I learned that there was this technology out there that could train a person how to achieve a certain state of mind through observed brain waves—which was technologically intriguing—and the potential seemed to be humongous. And the thing that seemed so out of whack was that I'd never heard of it."

Blue is an appealing salesman for his product—he's neither clinical-sounding nor New Agey. When I meet him, he's dressed in tan pants and a green short-sleeved shirt; he looks as though he might be an amiable local golf pro. "If you think of the world of neurofeedback as one that has been fermenting for many years, accumulating a lot of data that's really quite powerful," he explains, rolling out a gentle sales pitch, "we were looking to more or less uncork it."

I'm here to investigate something beyond the purely therapeutic application of the technology; I'm wondering whether a subculture of recreational neurofeedback users is on the rise. To date, neurofeedback machines have been largely clunky devices that look like they belong on the set of the '70s TV drama *Emergency*. The Attention Trainer system, on the other hand, looks like something you'd strap on to interact with your PlayStation. Make the technology sexy enough, and attach it to a Pentium IV, and people might start using it to expand their minds and not just to treat a specific disorder. Most recreational drugs began as medicines, after all—why couldn't neurofeedback follow the same course?

Blue is so impassioned in his description of therapeutic benefits that I worry I'll offend him by bringing up the recreational possibilities. "I'll talk to the kids after they're done," he says, staring me down intently, "and consistently what they say—it's like what one child said to me: 'Now I know how it needs to feel when I read.' They literally are learning what it feels like to pay attention. If you think about it—you're a little kid, and all you've been told is that you're not paying attention, you don't really know what the heck attention is."

I'm feeling a little sheepish now about even bringing up the idea of less clinical applications, but I plunge ahead, and immediately Blue's face brightens: "We think about those other directions all the time. This technology is not just for people who are battling disorders. It should be like going to the gym!"

Before long, I'm seated before a projection screen, while Kamran Fallahpour, a New York–based therapist who consults for the Attention Builders, adjusts the helmet to fit the contours of my skull. It feels vaguely viselike, but Blue is quick to remind me that it's designed to fit a child's head. After a few minutes of jostling, the computer reports a clean signal from my brain, and Fallahpour retreats six feet to the desk where the machine resides and clicks quickly through a few initial screens.

I'm listening to him speak, and as I'm listening, millions of neurons in my brain are building up tiny electric charges and then releasing neurotransmitters that indirectly send the voltage to other neurons via axons. These neurons communicate in unison, with a fantastically large number creating a collective rhythm with their discharges. That synchrony generates waves of electrical activity that can be measured by an electrode placed on the outside of the skull. About seventy-five years ago, a German scientist named Hans Berger discovered that the human brain generated a half dozen or so distinct wave states, each associated with a certain mode of consciousness: 1 to 4 hertz (or delta) appearing in non-

REM sleep, for example, or alpha, which lies between 8 and 12 hertz, and usually suggests a state of relaxation.

The sensors that are attached to my skull as Fallahpour speaks to me are capturing my theta levels, which fall between 4 and 7 hertz. Unusually high theta levels often accompany an easily distracted state of mind, so the software that Blue and his team have created is designed to reduce theta, pushing the brain into a more attentive state.

"Normally what we would do here is that we would have you play this memory game," Fallahpour explains, "and get a reading on your theta levels so that we could establish a baseline for the session. But I'll just play the game for you, since we're just doing a trial run here." I nod approvingly, not realizing the implications of what he's just said, and a few beats later, Blue jumps in to explain the need for a calibration session in the first place. He's building on his gym analogy from before, and so immediately he has my complete attention.

"If you think of this as being like exercise," he says while Fallahpour obligingly plays out the memory game on the screen, "the one thing that's different from, say, lifting weights . . . when you go to the gym, you're about as strong today as you were yesterday. But with attention training, depending on what you've been doing all day, you might be scattered or you might be quite focused. So you really need to establish that baseline at the beginning of each session; that way the rest of the session is tailored to your present state of mind."

The companion software application—one of the simplest in the Attention Trainer suite—involves an on-screen bicyclist who pedals faster as your theta levels decrease. The cyclist's exertion is relative to your theta levels during the calibration period. Think of your theta on a scale of 1 to 5—if you calibrate at 4 and then drive your theta down to 3 during the game, the bicyclist will pedal more rapidly. If you get it down to 1, you'll feel like Lance Armstrong.

My bicycle-pedaling skills, on the other hand, turn out to be more like Jabba the Hut's. After a few minutes of calibration, Fallahpour announces that the system is ready, and launches the bike game. A long stretch of intense humiliation begins.

From the very outset, my bike refuses to budge. I try focusing on my breathing. I try staring intently at the screen. I try staring intently at the wall. I try mentally reciting the first paragraph of a magazine article I've just written. The bicycle remains frozen. After about thirty seconds, the computer begins taunting me with pre-recorded encouraging words: "I know you can make it," it says in a voice that sounds uncannily like HAL refusing to open the pod bay doors in *2001: A Space Odyssey*. "Focus on the game."

I'm trying to focus on the game, but very quickly, I find myself focusing on the possibility that I have been suffering from ADD for years without realizing it. And then I find myself focusing on Blue and Fallahpour, who have settled into an awkward silence as I set what must be a new world record for theta levels.

After about a minute, Blue says, "You might try counting by sevens." I make my way up to 140, with no movement from the bike. Fallahpour says, "Stare at the cogs on the bicycle wheel, and concentrate on what they're doing." I fix my gaze on the cogs, which are spinning steadily despite the bike's overall intransigence. The staring does no good.

Five minutes pass, and the game ends. All told, the bike has made only a handful of brief nudges forward. I'm ready to swallow an entire bottle of Ritalin.

As I take the helmet off, Blue tries to sound upbeat: "You have to remember that this usually requires forty to fifty sessions for people to achieve their goals. It's hard to get results in one session." But I can tell he's a little puzzled by how poorly I've done. I ask weakly if it's possible that the system wasn't receiving a clean signal from the helmet. Everyone shakes their head—"You get an alert notice immediately if there's a problem with that," Blue explains.

And then Fallahpour pipes up, from behind the PC: "You know, there might have been something unusual about the calibration process. Normally, we would be capturing your theta levels as you play the memory game, but in this case we were capturing your theta levels as you were listening to Tom, and *I* was playing the memory game. Let's try it again, and do the calibration properly."

I strap the helmet on again, play the memory game dutifully, and we launch a new video game, this one with more advanced 3-D graphics. I'm using the computer keyboard now to drive a little go-cart through a cartoon cemetery—and the lower my theta levels get, the faster the cart moves on the screen. From the very first seconds of the game, I notice a dramatic difference: the cart charges ahead when I focus intently on the activity on the screen; when I move my eyes around the room, or flip mentally through a scattered series of thoughts, it lurches to a halt. I try counting by sevens, and sure enough, the cart picks up speed steadily.

After a few minutes, I ask to switch back to the bicycle game, just to verify the difference. After Fallahpour changes screens, the low-resolution cyclist reappears—only this time, I feel as though I can move him across the screen at will. He jumps forward when I concentrate, drops back when I lose focus. It's an absolutely uncanny feeling. I think a certain type of thought, and on the screen, something changes that reflects the nature of that thought. I find myself recalling the old Arthur C. Clarke line about the best technology being indistinguishable from magic, but this is better than magic. It feels like telepathy.

When I've finished with the demo, the company's head of product development, Ken Feldt, pulls up the behind-the-scenes data on my wave states during the session. The second, accurate calibration charted my theta levels at 3.6; during the subsequent games, I had reduced my levels to 2.7, thus enabling me to pedal the bicycle with much success. It turned out that during the initial calibration, when I had been listening to Tom Blue speak, my theta levels had dipped down as low as 1.6, twice as good as they had been during my most attentive moments pedaling—or failing to pedal—the bicycle. "That's why you were so bad at the game originally," Feldt says as he shows me the data. Without even trying, I had jumped into an extremely focused state in my conversation with Blue, more focused, in fact, than any of the focused states that I had deliberately attempted to reach. That state had set the benchmark for my first attempts to pedal the bicycle, and so when my theta levels during that initial session came in higher than my listening-to-Tom-Blue levels, the software responded by slowing the wheels to a standstill. It was my own version of "the zone." "We've found from our tests that some people can focus intently while they listen whereas others can't," Feldt says. "One of our clinicians was testing adults, and was working with a man who tested off the charts when he was reading but couldn't listen worth a damn. What this might imply about you is that you can focus very well when you're listening."

I found myself mulling over Feldt's words for days. After the immediate, breathtaking magic of the hardware itself—controlling a machine with my mind—it was this strange revelation at the very end of our session that stuck with me. Had I learned something new about myself? Maybe. I was more focused listening to someone talk than I was while actively trying to focus. I was more focused listening to someone talk than I was while locked into playing a video game. Of course, when I'm listening, I'm not thinking about my focus, which is probably one of the reasons I can focus in the first place. But it took the Attention Trainer's technology to prompt this glimmer of insight. I had caught a glimpse of what my brain was doing physically when it did something well. And all this was after just twenty minutes, and a slightly mangled calibration session.

* * *

A few weeks after my time with the Attention Builders, I met Leslie Seiden and Hal Rosenblum, a married Upper East Side couple in their late fifties and early sixties, who formed a company called Braincare, Inc., a few years ago. They make for an unlikely pair of neurofeedback evangelists—he's a retired pharmacologist and amateur photographer with a dry sense of humor; she's a lively practicing psychotherapist who has integrated neurofeedback into her practice.

When I visit them at their townhouse, she's wearing a bright pink suit with a gold brooch on her lapel. She's eager to tell her story. "I had been suffering from migraines since I was fifteen years old. I'd get them every few weeks, and that was just the story of my life. The medicines got better, but nothing really changed. And then one day I had dinner with a colleague of mine, and I was telling him about my migraines, and he said, 'Oh, I can help you with those.' And he told me about neurofeedback.

"So my husband and I went out and took a course," she continues. "I bought a used machine from a colleague and began doing sessions myself at night. I did about sixty sessions, and I really felt that it made some significant difference in my migraine history— they were shorter, less intense, less frequent. Also, I'd been a person who had to nap every day—after the neurofeedback, I stopped having to take naps.

"So I started to think: *this is pretty big.* I even got my kids to try it." Before long, they had converted the front room of their ground-floor office to headquarters of Braincare and put up a website advertising their services.

When Rosenblum hooks me up to Braincare's neurofeedback machine, the overall environment feels more medical than the Attention Trainer setup—Rosenblum has to attach electrodes to my skull with conducting paste—but once the software starts up, the experience feels familiar. It begins with a simple real-time portrait of my brain wave activity: four line graphs scrolling across the screen. Rosenblum points to one and says, "That's theta—that's what we'll be trying to reduce in this session." I stare at the data for a few minutes, as it rushes by, and I begin to experiment with different mental states. It's easier to throw the theta levels into overdrive by swooping from thought to thought than it is to focus intently, mainly because the data on-screen bustles with so much activity, and I find my eyes darting around the screen with each change. But even my attempts at artificial ADD don't last very long. There's a strangely hall-of-mirrors quality to my interaction with this machine: I try to act distracted, and within a few seconds, the wave form on the screen changes to reflect my distraction, which causes me to pay attention to it, thus ending my distracted state.

After a few minutes, I ask, "Do you have any good games?" And within thirty seconds, I'm piloting a spaceship hurtling toward a distant star, and once again I find that I can control the objects on the screen with ease.

In the four years they've been in business, Seiden and Rosenblum have treated nearly two hundred patients, mostly children battling ADD. But increasingly, the practice has attracted a more eclectic mix. Seiden tells stories of day traders trying to stay focused in front of a screenful of churning numbers, lawyers trying to reach "optimal mental functioning." "We had a Buddhist monk come in. He was in his sixties, and he'd lost his capacity for deep meditation," Seiden tells me. "We gave him ten sessions of alpha-theta, and he was able to find it again."

But even in the domain where it has had the most dramatic, if still anecdotal, success, neurofeedback has yet to become a mainstream approach. The sense of neurofeedback residing on the margins of popular acceptance is readily apparent to anyone exploring the field. In early 2002, I traveled down to a neurofeedback convention being held in a hotel just north of Miami. It was something of a surreal experience. If you didn't count the crowd gathered for the conference itself, the hotel clientele seemed entirely to comprise eighty-something retirees on vacation packages who boarded shuttle buses every morning for group expeditions to the Miami Boat Show or the Fairchild Gardens. While the shuffleboard set tottered over to buffet breakfast, a band of neurofeedback aficionados would be arguing over coffee and doughnuts about the merits of parietal lobe scans and phase portraits.

The group itself was a fascinating demographic: mostly boomers who seemed as though they had thoroughly enjoyed the '60s. There were a few Ph.D.s and a number of emissaries from institutes with exotic-sounding names. There were also at least two practicing parapsychologists. The mix of New Age sloganeering and neurotech jargon had, to my ears at least, a fresh, if not altogether persuasive, sound. "The problem with hearing both sides of the story," one speaker explained to robust applause, "is that you don't get to hear about all the other sides." The New Age component had me eyeing the exits on a number of occasions, but there was also an infectious enthusiasm to the group, in both their belief in the technology itself and their belief that they could use it to enhance their brains. It was a strange breed of the American ethos of self-improvement. I couldn't help being reminded of the personal computer hobbyists circa 1975: a high ratio of evangelists to ordinary users, united by a conviction that their technology could change the world. But there was a sense lingering in the air in Miami that the world had already taken a look at the technology, and turned up its nose. Some of these folks, after all, had been renegades in their particular disciplines for twenty years. "It's going to get worse," one of them announced when asked about mainstream acceptance of neurofeedback, "before it gets even worse."

Wes Sime stood out for me at the brain convention because in the midst of all the psi researchers and aging hippies, he got up on the stage and talked about golf. It was not out of character for Sime. The first time I talked with him, he was on a cell phone from a PGA Senior Tour event in Des Moines, Iowa, where he had been introducing professional golfers to the wonders of neurofeedback. A professor of health and human performance at the University of Nebraska at Lincoln, Sime may be the ultimate evangelist for neurofeedback's recreational potential. For one, his career has revolved increasingly around training athletes to use the technology, primarily using a Peak Achievement Trainer device made by a company called EEG Spectrum. And then there's his personal attachment to the technology, which seems almost obsessive. When I asked him over the phone to describe the gadget he's been using with the golfers, his voice perked up: "I'm wearing it right now—I've been wearing it throughout this conversation." A few days later, he sent me an email responding to a handful of questions I'd asked him, and his message ended: "Ironically, as I write this message to you I am using the neurofeedback software to shape my own attentiveness to the task. It is an interactive process that will soon become commonplace, almost like having a cruise control on my car."

Sime's first professional epiphany came from his work with a college diver who was recovering from a devastating back injury. After a series of neurofeedback sessions to work on his ability to focus—to imagine a successful dive before physically executing it— the young man made a spectacular recovery and was soon performing at a higher level than he had before his injury. "After he won his first meet, his coach came to me and said, 'I don't know what you were doing with Eric and all that head stuff, but Eric used to be able to dive, which meant that eight or nine times out of ten he'd do pretty well, but he'd always have one dive that would hurt him. But now— Now Eric is a diver. He makes something out of every single dive. I don't know how that happens—this is nearly unheard of because diving is such a precision sport.' Normally it takes a month to come back after an injury like that, but here this kid came back after only a week or two of training and won the first meet he was in."

Sime has also participated in a number of studies attempting to quantify these seemingly anecdotal accounts. In one remarkable project, dozens of golfers were hooked up to neurofeedback devices, and their brain wave activity was analyzed as they putted. Sime and his colleagues found a clear correlation between certain wave states and successful or unsuccessful putts. Armed with this data, the potential for "peak performance" training is obvious: once you know what wave states produce the most accurate shots, you simply need to configure your neurofeedback software to encourage that particular state. Interestingly, Sime says that the most effective setting for golfers tends to be an across-the-board, inhibit-all setting that discourages activity at all the major frequencies. It's the EEG equivalent of what athletes call "the zone": getting your mind out of the game and letting your muscle memory do its work unimpeded. For decades the zone has served as a kind of athlete's mysticism, but like Seiden's Buddhist monk, that mystical language is being translated into the hard data of science.

"I can literally go back with a golfer after a shot," Sime explains, "and say: 'Look at this, were you as focused as you wanted to be on the first half of this particular attempt at putting?' And he'll say yes or no, and I'll go down and look at the graph and say, 'Aha, you see here.' Or I can do it the other way around and say, 'You know what, on that last swing it looks like you started to bail out a little bit and started doubting your swing a little bit.' And the guy will go, 'You're absolutely right. I started to get anxious on the downswing.' It is the most exciting confirmation of quality of imagery and mental rehearsal that we've ever seen."

As Sime tells me this, I find myself thinking of the one time I saw Tiger Woods in person. It was the fourth round of the 1999 PGA Championship at Medinah, a tournament that he went on to win after a dramatic closing duel with Sergio Garcia. I was standing among the throngs lining the path between the sixth green and the seventh tee, reveling in the noise and the rhythmic chanting ("Tiger! Tiger!"), when the man himself walked down the narrow aisle carved through the crowd. For a second or two I saw him up close as he made his way to the seventh tee. I have never in my life seen a wider chasm between the look in someone's eyes and the surrounding environment. He had five hundred boisterous fans chanting his name from two feet away, and he looked as though he were halfway through a transcendental meditation session. If two people were cheering for me with such vigor, my heart would be beating like Secretariat's. Of course, Tiger Woods is used to crowd noise, but what I saw in his eyes that day was more than just the look of someone anesthetized to cheering. He had shut something off in his brain, and his eyes were reflecting it.

Shutting off instead of building up—Tiger's stare points to one area in which Tom Blue's mental gym analogy breaks down. There aren't many exercise machines out there designed to *weaken* muscles, or shut them down altogether. Sometimes training the brain is about learning how to turn *off* muscles that the brain naturally wants to flex. It's understandable that your brain wants to flush your body with adrenaline when five hundred people scream your name while swarming around you, but uncontrolled adrenaline may not be useful if you're trying to win your second major. So you learn how to shut it down. Athletes sometimes talk about "getting their brain out of the way," but, of course, you don't want your entire brain out of the way. An athlete wants to preserve the parts that take muscle memories and turn them into actual movements; he wants the high-speed corridors of instinctual movement to be active, while the more sluggish regions of introspection and self-doubt to be dampened. In a sense, great athletes are trying to reproduce the strategy that evolution stumbled across when it created the quick-and-dirty route that the fear response follows in the brain. If you don't have time to think, better to get rid of thinking altogether.

It's impossible to spend time in the world of neurofeedback enthusiasts and not feel intrigued by the vision of a personal neurotrainer who watches the computer screen as you practice your dives or your public speaking, rewarding you when you set your own personal best for theta inhibition. The idea might sound ludicrous, but it's not really all that far from what we're accustomed to today: every coach or teacher you've ever had—from Little League to college physics—was trying to condition your brain to behave in new ways. When you learned how to hold off on your natural impulse to swing at a changeup, or how to visualize travel at the speed of light, you were altering the neurochemistry of your brain: strengthening the connections between some synapses and weakening others; encouraging some broader regions to become more active while subduing others. The difference is that, unlike the neurotrainer, the Little League coach can't see the changes in brain activity directly.

Still, there are limits to the neurotrainer's vision. After the initial amazement had worn off while I was playing the space game at Seiden's office, I couldn't help noticing that I couldn't control the ship with nearly as much accuracy as I could have with a joystick or a keyboard. There's a fuzziness to the interaction that would be unpleasant were I actually interested in having an efficient conversation with the computer. Manipulating a computer with your brain is a bit like the old Dr. Johnson quote about the walking dog: it is not done well, but you are amazed to see that it is done at all.

That limitation has to do with the fact that the EEG sensors aren't terribly accurate.

I ask John Donoghue, executive director of Brown University's brain science program, what he thinks about these limitations. "There have been people who have been trying to get control of devices by picking up brain signals, but they've had little success," he explains. "Most of the experiments have been with paralyzed individuals: you hook up EEG equipment to them and put them in front of a computer. And then the EEG signal is fed into the computer in a way that might, say, make the cursor move up, choosing items from a selection.

"That's called one-dimensional selection, but the rate is pretty slow: like three words a minute. There are a few who can do two-dimensional control but it takes an immense amount of concentration. That's why, from the outside, using a noninvasive approach, it's pretty tough," he says. This is probably disappointing news for gamers who fantasize about mind-control versions of video games like Quake, but it also is a cautionary tale for those of us interested in the technology's capacity for enhancing introspection. The almost unthinkable complexity of the brain's information network is necessarily compressed down to a crude language when a machine listens to the collective rhythms of brain waves through the skull: hence my elemental interaction with the on-screen cyclist, consisting of two verbs—"faster" and "slow down."

"The information is there, but the problem is that getting access to it right now requires implanting something *inside* your brain," Donoghue says. "The only things you pick up with the EEG are big changes—that's why it's great for detecting epileptic seizures. You can track more global rhythms." In other words, at its best using neurofeedback to listen to your brain's activity is like hiring someone to attend a symphony for you who only reports back with word of each key change.

It was clear to me from these experiments that neurofeedback technology can accurately represent different mental states, albeit crudely. But it's still up for grabs how easily it can encourage brains to grow more comfortable in less familiar states. Systems like the Attention Trainers' or Braincare's had translated activity in my head into a new kind of language—that much was certain. But could these machines actually push my head in new directions? Given the climate of the last decade, the most compelling data on this question to date involves ADD. But the story is still incomplete. "I may have lost some credibility with some of my colleagues when I started offering this to patients," Seiden says to me. "But the data has been coming in."

Of course, if the technology does prove to be as effective as its evangelists believe it will, another fear quickly arises: are we going to create a generation of superattentive robots? A generation of kids trained on neurofeedback is probably preferable to a generation of superattentive stimulant addicts, which is where we seem to be headed now. And the Attention Builder software is open-ended. Already people are training themselves to reach alpha states, like Sciden's monk, or inhibit all, like Sime's golfers. Who knows—there might well be a high-theta subculture that emerges in the coming years. Some people read "Howl," some people read *The Seven Habits of Highly Effective People.* Some people inhibit theta, some encourage it. Neurofeedback is mostly just a mirror. How we choose to change ourselves based on what we see in the reflection is up to us.

But take the therapeutic question off the table, and you're left with the more provocative premise—that this is a technology that could lead to a more nuanced kind of self-awareness. The recreational use of neurofeedback could become an avenue for introspection, a way of bridging the physiological reality of your brain with your conscious mental life. We already accept the first two legs of Leslie Seiden's journey—from analyst to psychopharmacologist. If the "talking cure" and Prozac are now seen as legitimate avenues of self-exploration and improvement, why not a machine that listens to the sounds of our brains?

In the weeks following each of my neurofeedback sessions, I would catch myself stumbling into a certain mental state—early morning stupor, midtown Manhattan sidewalk anxiety, postcoffee email-writing fury—and wonder where my theta levels were, or if my beta waves were on the rise. I would think of Wes Sime, wearing his Peak Performance Trainer as he types out his email, and the question seems inescapable: could this become a kind of Walkman for the early twenty-first century? A Walkman that actually makes you faster, sharper, more in control. Assuming, of course, that faster, sharper, more in control is what you're looking for.

My journey into the world of neurofeedback made me more curious about what "attention" really meant. The more I thought about it, the more attention seemed like a colloquial illusion: it appeared to be a unified category only until you spent some time analyzing your own attentiveness. With enough scrutiny, the category began to fragment into component parts: there was me paying attention while playing the Attention Trainer video game; and there was me paying attention while listening to Tom Blue. They were both ways of paying attention, but when I thought about what the experiences actually *felt* like, they seemed to be two distinct activities, drawing upon different resources.

Playing video games tends to have a narrowing effect on my mind: there's nothing going on in my head other than a second-by-second appraisal of the activity on the screen. (It's a strangely trancelike state, despite the usual graphical pyrotechnics.) Paying attention to someone speaking, on the other hand, feels like a widening of my consciousness—or at least the part that grapples with meaning and semantic depth. The games are all about reflexes and reaction time; listening to speech is about interpretation—both the literal interpretation of the words and the mindreading assessment of the speaker's facial expressions, gestures, and intonation. The more I thought about it, grouping these two skills under the same umbrella seemed as artificial as saying that my juggling skills could reasonably predict my talents as a cook.

But if a single, all-encompassing category called "attention" was illusory, then what was a more accurate taxonomy? To say that one specific category is too broad is not to do away with categories altogether. Because of its importance in education, and because of the hype around ADD, attention turns out to be a much-analyzed faculty of the human mind. Even if the proverbial man on the street continues to think of attention as a unified thing, the neuroscientists and psychologists now know it to be a collection of different skills, sometimes overlapping and sometimes not. The concept of attention is a prisoner of our language: we think of these different skills as qualitatively alike because we have one word that embraces them all. Ultimately I realized I wanted to understand attention in the actual language of the brain, to learn about its core mechanisms. And then I wanted to test those mechanisms to find out more about my own attentiveness.

That's how I found my way to John Rodenbough, a North Carolina–based psychologist who has developed the Comprehensive Attention Battery (CAB), a software program that relies on more than a dozen separate tests to evaluate a person's repertoire of attention skills. The very first time we talked, it was clear to me that Rodenbough thought the idea of a unified attention category was misleading at best.

"People are often locked in to this idea that they have either good attention or poor attention," he explained in a quiet drawl.

"You'll often see these children who are labeled as having attention problems, but when you sit down and test them, you'll find that there are some areas that they excel in. I wonder whether there is such a thing as attention."

The most elemental distinction in the brain's attention circuitry is a relatively intuitive division among the different senses. You might have stellar visual focus, but you get easily distracted when listening to something or someone. Because they are easiest to test, seeing and hearing are the most widely studied of the attention faculties, but we also have olfactory and tactile attention circuits as well as "kinesthetic" ones that track our body's position in space.

Beyond sensory data, the component parts of attention revolve around how the information itself is processed in the brain. "Sustain" is your ability to remain focused on a single object or task for extended periods without becoming distracted. You might be great at sustaining olfactory attention, but your visual system might be easily diverted by new stimuli. At any given moment, so much data about the external world enters your brain through your sensory channels that the key proficiency of consciousness is not the ability to perceive the external world but rather the ability to shut so much of it out. If you paid constant attention to everything that your sensory organs were perceiving, you'd be overwhelmed with stimulus. Instead, the "mind's eye" focuses selectively on a tiny fraction of that incoming stream. The Danish writer Tor Norretranders calls this the "user illusion": you think being conscious means perceiving everything around you, but in fact it means perceiving small slices of reality and still being able to switch back and forth between them with extraordinary ease. That switching is essential to the illusion of consciousness, but it can lead to sustain problems as well. Suffering from poor sustain is like having a wandering mind's eye.

If sustain is all about remaining focused on an incoming data stream, "encoding" is the brain's ability to take that data and put it into working memory. The archetypal example is the encoding of phone numbers. To memorize a phone number, you first have to sustain your auditory focus long enough to actually hear the digits spoken to you; then you have to store those digits somewhere or else they'll be replaced by the next signal to come in through your ears. For short strings of data like phone numbers, the brain usually stores the information in what attention specialists call a "phonological loop"—like an audio recording of the numbers being uttered. This is often the case even if the original number was conveyed to you via sight. Watch yourself the next time you read a phone number off a piece of paper, and then walk upstairs to make the call. Most likely, as you read the numbers you'll repeat them to yourself—either out loud or internally—and then keep repeating the string as you walk up the stairs. You could theoretically memorize the shapes and arrangement of the digits on the page, and recall the number by recalling the image—but you don't. (We have innate skills as listeners, but our reading skills are all learned.) We do have powerful spatial memory systems, however, which is why we'll sometimes recall a number by capturing the spatial sequence of the numbers dialed on the keypad. But most of the time, you'll capture the number as a phonological loop. That process is what attention experts call "encoding."

Encoding is the attention subsystem that has received the most mainstream recognition over the past few years, largely because of the storage limits built into the human system. With very few exceptions, humans are capable of remembering about seven distinct items in working memory. (Technically, the figure is seven plus or minus two.) You can recall millions of separate things—everything from phone numbers to faces to the lyrics of "London Calling"—as long as they're stored in your *long-term* memory bank. But when new information comes along and you need to encode it quickly and retain it for a short while, you'll start overloading your working memory buffer if you go beyond seven items. It's no accident that phone numbers are seven digits long—when the phone companies set out to design the modern dial system, they consulted psychologists on the maximum number of digits that could be readily memorized by an average person.

After sustain and encoding, the toolbox of attention grows more complicated, because attention isn't just about focusing on a single task or object—it's often about switching among different tasks and different sensory inputs. One measure of this is what specialists call "focus/execute" skills. Assuming you don't live in a monastery or a prison cell, every day of your life probably involves running through thousands of separate routines that follow a regulated script, with each stage requiring a specific attentional mode. You check the kids' seat belts, put the key in the ignition, listen to the engine, glance over your shoulder to make sure the driveway's clear, look both ways before you turn onto the street. If this is a familiar sequence for you, odds are you perform these actions almost unconsciously by now— but not entirely unconsciously. If something goes wrong at any of the stages—if you see an oncoming car or Junior's unbuckled his seat belt—you'll notice, because on some fundamental level you're paying attention.

Attending to all those details would be overwhelming if each wasn't finite in duration and objective. You can run through the routines like clockwork because at each step your brain knows to stop the previous task and start the next one. If your brain weren't capable of making these shifts, the incoming data would quickly add up to information overload. You might not notice that skateboard in the driveway because you were still thinking about the ignition key. "Focus/execute" describes the sequence when it's done right: you focus on a specific task, execute it, move on to the next task, and focus all over again. Rinse, repeat.

Focus/execute implies a predetermined script, but of course real life doesn't always supply a script. Our most sophisticated feats of attention usually come when we have to extract on-the-fly assessments of relevance out of a flurry of competing signals. This is attention's executive branch, which usually goes by the name "supervisory attention control." Supervisory attention is the quarterback who sees the open receiver thirty yards away despite the linebackers hurtling toward him; it's the music aficionado who can pick out a single ill-tuned violin lurking in an orchestra's vast sound field. (Or it's the parent noticing the toy wagon left in the driveway as she's backing out of the garage with three kids cackling in the rear seat.) People with supervisory attention skills are often good at shutting out stimuli that we should naturally be attentive to—Tiger Woods blocking out the five hundred cheering fans as he walks to the seventh tee. In this sense, supervisory attention often involves overriding our impulses, voting down the obvious attraction in favor of a more understated object of study.

The attention system works as a kind of assembly line: higher-level functions are built on top of the lower-level functions. So if you have problems encoding, you'll almost certainly have problems with supervisory attention. When people notice attention impairments, they're usually detecting problems with the focus/execute or the supervisory levels, but the original source of the problem may well be farther down the chain, or it might be localized to a particular sensory channel. For a psychologist like Rodenbough, the first step in treating an attention disorder is to isolate the weak link in the chain. That's why he developed his CAB software—a suite of distinct tests tailored to measure the components of the attention system. The CAB tests don't peer inside your brain directly, of course, but they are ingeniously designed to detect both the strengths and weaknesses of each tool in the attention toolbox.

* * *

It's fair to say that taking the Comprehensive Attention Battery is like playing the worst video game you've ever played in your life. When I first sat down to explore the CAB tests, I tried to pump myself up: you get a score at the end of this test, after all, and I wanted a high one. This would be my own private attention decathlon—all I had to do was stay focused for an hour or so, and my extraordinary gifts would be recorded for posterity.

Then I started up the software, and before long, my brain was hurting. Taking the CAB was a kind of karmic payback for all the times that I found it amusing to recite random integers within earshot of someone trying to memorize a phone number. Because each test is contoured to probe the brain's attention system, it forces you to confront the limits of your specific attention faculties. As the tests unfold, they tend to get more difficult; invariably there's an inflection point at which you can feel your hardware shorting out. But it's not a general lack of focus that you experience; it's an extremely precise sensation—the difference between your car feeling a little slushy while making turns and a light popping up on the dashboard alerting you that the pressure is low in the front left tire. If you've ever doubted the "law of seven," you should try taking the audio encoding test. It's the most elemental of the tests: the computer begins by listing three numbers, which you have to enter back in the correct sequence after a momentary pause. The process is effortless for the first few rounds, as you need to encode four then five then six numbers. Encoding six numbers is just as easy as encoding three—you can play back the phonological loop as though it were an audio sample reliably triggered by a keyboard. The information doesn't degrade. But once you cross over into encoding eight or nine numbers, your brain starts to scramble. You can feel the last few digits pushing the first few out of the buffer.

As your encoding system approaches its limits, it instinctively reaches for shortcuts, reducing a collection of digits to a single data point, thereby freeing up room in your working memory. As I neared the end of the encoding test, one ten-digit number I was asked to memorize began with 3–0–1, my parents' area code. Instantly, I was able to translate those three numbers into a single unit, leaving more room for the rest of the sequence. Instead of encoding ten random numbers, I had to encode seven numbers plus my parents' area code. Eight items, instead of ten—just few enough to keep the whole sequence clear in my head. Memory specialists call this technique "chunking"—turning a series of discrete objects into a larger chunk, freeing up working memory for additional data. (Four-digit numbers that begin with 19 are particularly vulnerable to being chunked, since they can be memorized as years—instead of encoding ten numbers, you encode six plus the year JFK was shot.)

The CAB illuminates a few quirks of the brain's attention architecture. The encoding tests include both forward and backward versions. For the latter, you have to memorize a series of digits and then enter them in reverse order. This part of the test had me reaching for the Advil even before I passed the seven-digit threshold. Having to punch the numbers back in reverse order meant that the phonological loop wasn't nearly as helpful as it had been in the forward version. You could play the mental tape of the sequence, but there wasn't a ready-made internal mechanism for playing it backward. I got only three out of twenty wrong in the forward version, but twice as many foiled me going in reverse. That disparity made intuitive sense to me: having to reverse the sequence felt like something my brain wasn't designed to do. But when I moved on to the visual encoding test, I stumbled across a startling result. The test shows you a grid of nine boxes, arranged tic-tac-toe style. Instead of listing a series of numbers, the test illuminates a sequence of boxes. After the sequence is over, you repeat the sequence by clicking on the boxes: first in the original order, and then in reverse. Unlike in the audio test, visual encoding going

backward was *easier* than going forward. The sequence came naturally to me in reverse order, while re-creating the original series
took more effort.

After the test, I ask Rodenbough if my results with visual
encoding are unusual. "Not at all," he explains. "Our brains are
designed to track backwards visually. When you see some kind of
movement—our brains are designed to follow that movement
backwards." When you track a projectile flying through the air,
your brain intuitively calculates its point of origin by imagining its
trajectory in reverse. It's one of those little talents you've relied on
your entire life without ever really noticing it. But taking the CAB
made that particular faculty as vivid as those hardwired for stereo
vision or face recognition.

Reverse visual encoding is a general human aptitude. Some of us
are better at it than others, thanks to our genes or our cultural
training—but on average, we're all better going backward than forward while capturing visual data. It's a species trait. Taking the
CAB helped me recognize that I possessed the trait, and that itself
was a kind of insight. But I was searching for individual variation
as well—not just the universal attention toolbox, but also my own
distinct talents with the tools themselves.

I ask Rodenbough if the tests have caused him to think about
his own attention faculties any differently. At first, he demurs. "I
have to remind myself that my scores on the tests themselves are
meaningless. I've spent so many hours debugging the software that
I must have taken each of them hundreds of times. Over time,
though," he concedes, "I've learned that I don't do very well sustaining attention auditorily. I tend to get a bunch of thoughts in my
head, and I lose track of what someone's saying."

I ask him if knowing about this deficit has changed his attention
strategies in noticeable ways. "Well, my wife complains that I don't
listen to her," he says after a pause. "So I try to think about this in

terms of the components of attention. Am I listening to her, or am I thinking about too many things at the same time, and she's just running over my buffer for my encoding ability? That's how I've been able to rationalize it—when she says something, I'm thinking over all the permutations of what she's said, and it uses up all my encoding space."

"So it's as though you're listening *too* intently?" I ask, a smile forming.

"That's right."

"And does she buy that?"

"Well, no, I haven't given her that one yet." Rodenbough laughs. "I just keep it to myself."

The most intriguing tests in the CAB are those that engage the mind's executive branch. Taking the encoding and sustain tests makes you feel as though you're pushing the limits of a cognitive muscle that you don't really have much control over: you hit eight digits on the encoding, and no matter how hard you concentrate, you can't keep the entire string in your head. But as the tests start to explore your supervisory attention skills—the ones atop the focus food chain—the part of your head that feels like "you" starts to come into play. I found these executive branch tests revealing because they came closest to capturing the real-world experience of trying to "pay attention," particularly in an age of mixed media and sensory overload.

Supervisory attention is ultimately all about choice: your executive brain receives all sorts of data at once, pouring in through the various sensory channels, and somehow you have to decide what's important and what's not. The CAB tests present incoming stimuli as fairly elemental components, but then they go to maddening extremes to cross the signals between the different channels. The most famous example of this is the Stroop Interference Test, in which three words—"blue," "red," and "green"—repeat randomly in any of the sixteen squares of a grid. The words themselves are colored blue, red, and green, but not consistently: sometimes the word "red" is colored red, but sometimes it's blue. In the initial stage of the test, your job is to select all the words that match the ink they're printed in, all the red reds and blue blues and green greens. This is much harder than it sounds, precisely because your brain has to resolve conflicting information coming in from different senses. As your attention settles on each word in the grid, a strange duet develops in your head: *I know the letters spell out "blue," but is the color blue?* As you stare at the word, the letters insistently bark out blueness to the part of your brain that processes language, but your visual system is making a very different report: "What do you mean, blue? Those letters are red!" Part of your brain sees blue, and part of it sees red, and your executive brain has to make the call.

When I took the Stroop test, I found myself dealing with the conflict between these two modules by shutting down my language processor as best I could. I tried to see each word as a pure shape and not a series of recognizable letters. It helped that each of three words was built out of a different number of letters, giving each a distinctive width. And so—without even truly realizing what I was doing—I found myself scanning the grids looking for small blocks of text colored red, medium-sized text colored blue, and lengthier blocks colored green. I was feeling pretty proud of myself until I got to the second stage, at which point a recorded voice began chanting "red, red, blue, green, green" as I tried to pick out the correct squares. It was about then that I started crying.

My tour through the CAB may not have been the most fun I've ever had at a computer, but it left me with an oddly precise awareness of my different tools as I utilized them in real life. In the days that followed, I'd memorize a phone number and think, *Right, this is auditory encoding.* Or I'd switch back and forth between watching CNN and reading my email and think, *This is supervisory multiprocessing.* Before I would simply have said that in each case I was trying to pay attention. Now the two acts seemed as different as push-ups and running on a treadmill. They exercised their own cognitive muscles, and taking the CAB had allowed me to perceive those muscles as distinct entities for the first time.

Unlike Rodenbough, I found that my visual encoding—for faces and environmental details—was the weakest link in my attention chain. Having isolated this property by taking the test, I started to confirm it in ordinary life. Around the time I was investigating the CAB, my wife and I were in the middle of a complicated renovation of a new house we'd bought. We'd take trips out to inspect the progress, and on return, my wife's brain would be filled with dozens of seemingly photo-realistic details from the house, while mine would have a few meager scraps of images and general impressions. We had looked at the same objects, but I had failed to encode them. I started to think about it in the language of computer software: my default settings include visual encoding turned off. For my wife, I suspect, it's the opposite: just walking around a room fills her head with details that she can recall days later. This doesn't mean that I'm not capable of remembering visual information. In fact, now that I've located the problem, I've improved a little by consciously switching on the encoding routine when I'm in an environment I want to remember. Instead of scanning a room passively, I break it down into component parts: "Okay, notice the molding over the doorway—there's a crack there. Now look at the electrical panel here. . . ." It still doesn't compete with my wife's skills, but at least I'm in the game now.

After my sojourn with the CAB, I experienced a strange, no doubt illusory side effect. Because I had learned a little about brain anatomy, I started to feel as though the different modes of attention were emanating from different physical locations in my head: the supervisory tasks were localized around my frontal lobes while the more primitive tasks—like sustain—seemed be happening in the back of my skull, closer to where incoming visual data is processed. All the brain scientists with whom I have discussed this claim that this kind of intracranial spatializing is impossible: you can't literally *feel* where the computation is being performed in your head. But in a way, my having concocted the sensation was even more telling: as a species, we're great mapmakers; our intelligence naturally gravitates to spatial organization. (One theory holds that the seat of long-term memory in the brain, the hippocampus, originally evolved as a cognitive mapmaking tool, helping our ancestors get their bearings in complex natural environments.) I had mapped in my head the brain's attention system, and I had a newfound sensitivity to the specific components of that system, so it was almost inevitable that my brain would layer that map over itself, like a trick of the mind's eye.

Jettisoning the idea of attention as a Single Unified Thing leaves you with two primary implications. The first we've already seen: if the art of paying attention is actually divided among several different modes, it's helpful to learn which of those modes works well for you and which ones don't pull their weight. The second insight operates one level up: if your attention is a system of interacting modes, then one of the most essential high-level functions that your brain performs is switching those modes. You can be the most brilliant auditory encoder in the world, but if you can't switch into auditory encoding mode when it's appropriate, your talents will be wasted. Part of having an effective brain is possessing good tools, but an equally important part is being able to pull the right tool out at the right time.

Shortly after I'd spent my time with the CAB, I traveled west for my last expedition into the world of neurofeedback—to the offices of the Othmer Institute, located on the other side of the Hollywood hills, not ten minutes from that icon of Valley girl culture, the Sherman Oaks Mall. The institute is run by a husband-and-wife team, Susan and Sigfried Othmer, both practicing psychologists and longtime neurofeedback advocates. Sue had agreed to talk a little about their practice, and do some training with me. Driving out to their offices, I had been thinking again about the importance of mode switching, and so when Sue and I finally sit down in her office, I bring up the idea early in our conversation.

Before I even finish my sentence, Othmer nods emphatically. "We look at everything we do as improving brain self-regulation," she explains in a quiet, confident voice, refreshingly free of the evangelical overtones I'd encountered among other neurofeedback practitioners. "Our states vary enormously over the course of twenty-four hours, but we don't perceive it that way because usually the state is appropriate for the moment." Tools like the Attention Trainer were designed to push your brain toward a single target, but Othmer was more interested in exploring different states: focused but calm, or high-energy, or meditative trance. Because the software is artificially propelling your brain into these modes without altering your surrounding environment, you perceive the states themselves with a newfound clarity.

"With neurofeedback, it can be very odd to suddenly feel yourself in a different state, for no apparent reason," Othmer says with a chuckle. You don't notice what the daydreaming state feels like normally, because you're too busy daydreaming. But when you drop suddenly into daydreaming mode with a set of electrodes pasted onto your skull and a total stranger sitting two feet away from you, the shift in focus pops out immediately. "If I train you too low, you'll feel a little stoned, a little drowsy—you might not want to drive," Othmer says as she starts up the computer for a demonstration. "If I train you too high, you'll be bouncing around the room."

Othmer begins the training by showing me a recording of someone else's brain waves. On the screen, there are three scrolling lines, each representing a different part of the frequency spectrum. Sharp spikes appear at regular intervals in the top line. Othmer points to them: "You'd normally see those in a drowsy state. But this guy, he's actually wide awake while we're recording this. So you can see he has a severe attentional problem." As she's talking, she's placing the electrodes against my skull. After a few minutes of fiddling, she hits a button on the screen, where a single scrolling line appears.

"This is you," she says. And there I am, or at least some small part of me, whittled down to a jagged—though thankfully not spiky—line on a computer monitor.

"I'm going to take the brain waves, and then break out different frequencies," she continues. With one click, the single line becomes three. "Then I'm going to set thresholds on some of those frequencies. I'm going to reward you every time you increase the amplitude within those thresholds." The visual feedback on the monitor makes Othmer's description immediately clear to me. As the wave form shuttles across the screen, bouncing up and down, Othmer creates two boundary lines, above and below the wave. Making those boundary lines closer to each other decreases the room available for the wave form, while pushing them farther apart opens up more room. The goal of the exercise for me is to fill the space between the lines as much as possible, without going over the boundaries.

By changing those thresholds, she can indirectly change my internal states. "So we have a hook, and we can grab the state and move it up or down. Lower the rhythm, and the state becomes deeper; make the rhythm higher and it becomes more active," she says.

Of course, all this depends on my ability to change my internal state to match the changing thresholds. That's where the reward comes in. I ask her how I'm to be motivated—I'm thinking candy, perhaps, or gold stars—and she switches on a second monitor. A streamlined version of Pac-Man appears on the screen: a maze of white dots with a circular creature poised in the upper left-hand corner, ready to explore. "This is your reward: when you increase the amplitude, the Pac-Man will start moving through the maze, and you'll hear a beeping. The process is pretty much 'let it beep, and be pleased that it's beeping.'" I tell her that this sounds right up my alley, and she laughs. "It's usually really easy for kids and almost impossible for adults the first time around."

Othmer suggests that we start with a more active, alert state. She hits a few buttons, and the session begins. I stare at the Pac-Man and wait a few seconds. Nothing happens. I try altering my mental state, but mostly I feel as though I'm altering my facial expression to convey a sense of active alertness, as though I'm sitting in the front row of a college lecture preening for the professor. After a few seconds, the Pac-Man moves a few inches forward, and the machine emits a couple of beeps. I don't really feel any different, but I remember Othmer's mantra—"be pleased that it's beeping"—and so I try to shut down the part of my brain that's focused on its own activity, and sure enough the beeping starts up again. The Pac-Man embarks on an extended stroll through the maze. I am pleased.

After I've made it all the way through the maze, Othmer asks me how I feel. I do a quick internal check, and report that I indeed feel a little more alert. Not quite caffeinated, exactly—more like that upbeat anticipatory state as I'm watching the coffee being poured. Othmer offers to take me down a few notches, and I begin my mental dance with the Pac-Man once again. This time, I find the best way to get the little sprite to move is by mimicking the zoned-out state I often experience over a bowl of cereal first thing in the morning—in other words, your classic "spacing out." It takes me a few minutes to drive the Pac-Man through the maze, and when I'm done, the "down" state lingers, not unpleasantly.

Entering that zoned-out state so readily makes more of an impression on me than the alert one, partially because it brings into sharp relief that state I've been in since I walked in the door. Having a conversation with a new person—particularly in an interview setting, where there's a rapid segue from the usual chitchat about the weather to more important ideas—invariably puts me a little on edge. I find myself talking faster, in both my exterior and interior monologues. On the outside, I make a lot of jokes, while on the inside, my mindreading usually kicks into overdrive. ("Does she think I'm an idiot? Why do I keep making these jokes?") So when Othmer tried to amp me up with the neurofeedback, the change wasn't all that noticeable. But finding myself in that slower, deeper state after the second training came as a shock. I'd only met this person twenty minutes before, and yet there was no trace left of my initial-meeting buzz. I thought to myself: *I wouldn't mind learning how to downshift into this state on command, if the setting warranted it.* This seemed to me the antidote to the fears of a neurofeedback-enhanced generation of superattentive people: you could use the technology to take the edges off or you could use it to make the edges sharper. You could attempt to improve focus, or you could learn how to make things a little blurrier. And maybe more important, you could use the technology to help you select the appropriate state on command.

Until I met with Othmer, I had thought of neurofeedback as a tool-sharpening device rather than a technique for improving your ability to shift between different tools. But Susan and her beeping

Pac-Man helped me see that changing modes was a skill in itself, and one that you could be trained to perform better. There are modes, and then there's mode switching. Both areas are essential to learning how to use your brain. "Mode switching isn't something you're taught at school," Othmer tells me after the session. "But that's what we're trying to teach here."

4

Survival of the Ticklish

Like many first-time parents, within minutes of learning that my wife was carrying our child, I started calculating ways that the pregnancy could go wrong. As the months went by, I was haunted by the usual worries: the risk for miscarriage, a first-trimester bout of food poisoning, bizarre appendages in the ultrasound. Then in the sixth month, the baby dropped unexpectedly, and our obstetrician suggested that my wife go on bed rest to prevent a premature birth, where she remained until the final few weeks. As we grappled with that confinement, my mind piled on additional anxieties, ones that probably would have been better suited for a sitcom. Our doctor's hospital was located on the Upper East Side—a twenty-minute cab ride from our apartment with no traffic, but counting on no traffic in midtown Manhattan is like counting on a New York subway rider giving up a seat for a pregnant lady. I had visions of delivering the baby myself in the back of a cab, somewhere on the shoulder of the FDR Drive, with a squeegee guy as the mid-wife.

But somehow our son managed to remain in utero until the due date, and even waited out the cab ride to the hospital. He was born with no complications, and after two days at the hospital, mother and child were released with a clean bill of health. When we made it back downtown to our apartment, I started to notice the profound lifting of anxiety that had begun forty-eight hours before, when our cab pulled up in front of the hospital and it became clear that my fifteen minutes as an amateur ob-gyn would have to wait. As I went to sleep that night, the three of us together in the same room for the first time, I realized how much background anxiety I had been carrying around for the past nine months—and how little I now felt, with my wife and child sleeping contentedly beside me. I thought to myself: *I can't wait to get up tomorrow and experience what it's like to walk around so free of fear.*

But of course things didn't work out quite as I had imagined. Because tomorrow turned out to be September 11, 2001.

My memories of that horrific day are mostly the now-familiar ones: seeing the towers burn from twenty blocks away, watching them fall on TV, feeling that sense of dread as the networks reported that another plane was missing. But a particular memory stands out, and in the days and weeks that followed I found myself puzzling over what it meant. That memory is of standing next to my wife, as she nursed our three-day-old son in a rocking chair, telling her that the second tower had fallen, and seeing the strangely distant, becalmed look in her eyes. I was filled with a new father's protective fretfulness: were we safe in our apartment? Should we leave the city? Would the air be dangerous for our son? But my wife looked almost as though I were describing a fender bender I'd witnessed on the way home from the supermarket.

Later, she confessed that she had found the whole experience otherworldly—while she felt guilty about not being adequately affected, she couldn't bring herself to experience the shock and terror that everyone else seemed to be going through. Logically, she understood that something terrible had happened, but she couldn't *feel* it. It wasn't as if she were indifferent to the potential threats: she advised the rest of us in the house to do all the right things—stock up on bottled water, call the pediatrician for advice about the air-quality issues. But you could see in her eyes, in her whole carriage, that the crisis wasn't having the same effect on her as it was on others.

Something in this seemed odd to me: I would have expected a new mother to feel an exaggerated sense of alarm, given the tiny life she was cradling in her arms. Wouldn't the instinct to protect her child trigger a *stronger* fear response? I could feel the adrenaline coursing through my body that day as we contemplated our options, but my wife seemed like she'd just taken a Valium. What was going on?

The answer, I learned later, was that we were each on very different drugs. While I was under the influence of the cocktail of hormones that creates the fight-or-flight response, my wife was being lulled by a very different chemical—a fascinating molecule called oxytocin that plays an essential role in some of life's most profound emotional events: falling in love, forming strong social attachments, having a baby. As I paced furiously around our apartment glued to the latest CNN update, oxytocin was keeping my wife calm and nurturing. And though I had my son's best interests in mind, I have a pretty good idea which of our responses was the more helpful at the time.

A year later, I found myself on UCLA's Westwood campus, meeting with Shelley Taylor, a psychology professor who has done extensive investigation of the relationship between oxytocin and stress. Taylor is one of a growing cadre of brain researchers who have placed a newfound emphasis on the "positive" emotional circuits in the brain. For a number of reasons, the brain sciences historically have spent far more time exploring the neural pathways of negative emotional responses: on today's map of the mind, the regions of fear and stress are clearly delineated, with mostly minor border disputes remaining. Until recently, the kingdom of love and affiliation was a vast terra incognita, so thinly sketched that you almost didn't notice its absence. But Taylor and her allies have started to change all of that.

When we meet in her office, a vista of downtown L.A. hovering behind the pine branches outside her window, I begin by asking how she got interested in oxytocin (not to be confused with the oft-abused painkiller oxycontin). Taylor tells the story of attending a lecture in the late '90s given by a visiting scholar on the subject of stress and the fight-or-flight instinct. At one point in the lecture, the speaker discussed the levels of aggression displayed by his lab rats when they were exposed to stress. After exposure to regular stressors such as electric shocks, the animals would literally bite and claw each other to death if they weren't separated.

"That went off like a lightbulb in my head, because it's not at all descriptive of what we typically see in our human studies," Taylor tells me. "So I went back to my lab group and I said, 'What do you make of these disjunctions between the animal studies and what we see in humans?' And one of them said, 'You know, the animal studies are all based on males. They don't include females at all, because females cycle so rapidly.' And then someone else said, 'You know, I think that's true for the human literature as well.' So we started looking through the literature to see how well female responses to stress were represented, and the answer was very poorly. Prior to 1995, they constituted seventeen percent of the participants. There were virtually no studies where you had enough female participation to do a comparative study."

This lack of gender parity was not just a political issue. For decades, the scientific literature on stress response revolved around a fundamental causal chain: introduce a stressor—a lunging predator, say, or a rival stealing your food supply—and the body initiates the now famous fight-or-flight response. Taylor suspected that the fight-or-flight instinct was only half the story: "I said to my group, 'Okay, let's start from scratch. What are women doing? Is fight-or-flight a reasonable description of women's response to stress?' And within seconds, all of us had an immediate response: 'No.' Because what differentiated female responses to stress from males' is that female responses have to incorporate the protection of offspring, at least for the period of time that there are offspring. Our idea was that fight behavior works fine if you're an individual; but if you're trying to protect [your] young, fighting just isn't going to work. The same goes for flight—only ungulates like deer have offspring that are capable of fleeing shortly after birth."

Two years after attending the original stress lecture, Taylor had formulated her response, in the form of an essay published in *Science* called "The Tending Instinct." Fight-or-flight was one way of dealing with stress, she argued, but there was another option: tend-and-befriend. You can combat stress by literally going into combat with it, or you can reach out to your support group. Both stories are integral to the human experience, although Taylor believes that the tending instinct is more commonly expressed in women. She says, "There was recently a metareview of twenty-eight different studies, and twenty-six of them found that women sought social support in stressful situations more than men. Short of childbirth, there is no sex difference in humans that looks like that. With most sex differences—men have a slight spatial advantage, women have a linguistic advantage—when you actually look at the curves, there's an enormous amount of overlap." But when it came to seeking out social bonds in the face of stress, the data was emphatic.

Taylor and her team even had a solid hunch about the brain chemistry behind the tending instinct, and oxytocin was a central player. Researchers had long known that oxytocin was released during life experiences that involve intense emotional attachment: childbirth, breast-feeding, and sexual climax. Higher oxytocin levels had been linked to sheerly stressful experiences as well. While oxytocin is present in both male and female brains, evidence suggested that estrogen enhanced the hormone's effects, making it less powerful in testosterone-heavy male bodies. If there was a biologically grounded tending instinct, oxytocin probably played a role in it.

You can't dig very far into the literature on oxytocin without encountering a memorable little creature that to date has shed more light on the neurochemistry of attachment than any other animal. The prairie vole, a small rodent indigenous to the midwestern plains of the United States, is one of the natural world's great romantics. After mating, most voles remain monogamously attached to their partner for life, raising children together in a rodent version of domestic bliss. This is, to say the least, an unusual practice in nature: among mammals, only 5 percent show this sort of monogamous, biparental behavior. Around twenty years ago, a neuroendocrinologist named Sue Carter began examining the brain of the prairie vole in an attempt to understand what caused its unusual fidelity.

"I became interested in oxytocin then because I knew that oxytocin was released during sexual behavior," Carter tells me over the phone from her office at the University of Wisconsin at Madison. "There was already research coming out showing that oxytocin facilitated parent-child bonding in sheep." When Carter injected oxytocin into the brains of the voles, the animals formed even more tenacious bonds than usual. Carter also explored the effects of oxytocin from the reverse angle by injecting chemicals that shut off oxytocin receptors, blocking the hormone's effects. Instantly, the prairie voles' lifestyle became less *Leave It to Beaver* and more

Woodstock: indiscriminate mating without any lasting attachment. "The most compelling evidence for oxytocin's role in bonding is simply that when you block the oxytocin receptors, the animals don't form pair bonds," Carter explains.

Several years later, an Emory University professor named Tom Insel, now president of NIMH, began a comparative study analyzing the brains of prairie voles and their less monogamous cousins, the montane voles. Insel discovered a remarkable difference between the two species: in the faithful voles, the oxytocin receptors overlapped with dopamine receptors in an area of the brain called the "nucleus accumbens"; in the nonmonogamous voles, the oxytocin receptors were located elsewhere. The nucleus accumbens is generally regarded as one of the brain's essential pleasure centers, while dopamine coordinates many seeking and appetitive behaviors. In the monogamous voles, in other words, oxytocin receptors were planted firmly in the reward circuitry of the brain. The architecture suggested that behaviors associated with oxytocin release would feel good in the brains of the prairie voles, but have a different effect on the montane voles. If oxytocin encouraged the animals to stay attached to a partner, it was no wonder the prairie voles turned out to be so committed. Their brains were wired to make forming attachments pleasurable.

The temptation to translate the vole studies into the brain chemistry of humans was irresistible. Like those of the monogamous prairie vole, human oxytocin receptors are located in several dopamine-rich regions of the brain, suggesting that oxytocin is also embedded in the human brain's appetite and reward circuitry. One study compared the brain activity of subjects looking at pictures of loved ones versus pictures of nonromantic friends. The pattern of activity in the cortex was markedly different depending on which type of face the subject was exposed to. Interestingly, fMRI scans of the brain processing a romantic gaze bear a striking resemblance to the brain activity of new mothers listening to infant cries. They also resemble brain images of subjects under the influence of cocaine. (We'll return to this last similarity later.)

The face recognition studies are of particular interest because a number of animal studies have convincingly linked oxytocin to the formation of social memory. One hypothesis is that oxytocin release during key pair-bonding events like sexual climax or childbirth helps cement the image of a partner or a newborn in the mind's eye. Mothers who breast-fed their children often describe powerful memories of their infant gazing up at them during nursing. The vividness of that memory, and its association with warm, maternal feelings, may well be the imprint of oxytocin.

There is something about the idea of a dedicated love circuitry in the brain that rubs certain people the wrong way. We accept readily enough the idea that our fear response should have its own chemical and neuronal architecture, but somehow it seems demeaning to suggest that a comparable physiological substrate exists for feelings as rich as love. Once over dinner, I told a friend of mine a few tidbits about oxytocin and the neuroscience of attachment. He'd sat through a number of other brain-related riffs with enthusiasm, but when I moved on to love, he gave me a suspicious look: "I have a hard time believing that there's that much commonality in the way that people experience love. I bet if you did scans of people's brain in the throes of romantic love, it would look different every time."

To a certain extent, my friend was right: our brains are like fingerprints, each one a little different. When you think of a loved one, a unique constellation of neurons fire, triggering the image of their face, memories of past times together—a subtle cocktail of different subemotions. Some people no doubt experience love more vividly; some find it linked inextricably to sexual attraction, while others have more platonic inclinations. In theory, all these differences could be detected by advanced brain scans.

And yet beneath all that buzzing diversity, a few core patterns remain. By definition, emotions this fundamental need an underlying circuitry to do their work. If love lacked a physiological basis—if it were just something we decided to pick up, like learning how to type or play the cello—the emotion wouldn't have the transformative, and sometimes destructive, power that it wields over our lives. Part of understanding love is celebrating the differences, but another part is sharing in the common experience. That's why romantic poetry works, after all.

In a way, it gets down to the question of what you mean when you talk about a "unique" profile. Every fingerprint is unique from one angle, in that its telltale markings differentiate it from every other fingerprint on the planet. But from another angle, all fingerprints are the same: grooves in our front skin arranged in semiconcentric circles, with a reliable series of components: center points, detch points, delta points. Love is like those fingerprints: the component parts are invariably arranged in novel ways, but the components themselves are universal.

The complexity of the human brain—and the ethical problems of experimenting with love research—may mean that the scientific understanding of human attachment may not proceed as quickly as some would hope, which will probably come as a relief to the romantic poets among us. But while our knowledge of human neurochemistry is finite, the extent to which the chemistry repeats itself in other mammals suggests that love is as much a part of our evolutionary heritage as heartbeat regulation or stereo vision. If we had evolved as a species with different mating and child-rearing habits—abandoning our children at birth and moving indiscriminately from partner to partner, as most reptiles do—it's likely that our brains would be incapable of feeling love.

Reptiles lack both our neocortex, the seat of language and higher learning, and most of the human brain's limbic system, which, as we've seen, plays a key role in regulating emotional response. Reptile brains do not produce the oxytocin molecule at all. If some accident of evolution had led reptiles to develop advanced forebrains capable of language and higher-level consciousness, while maintaining their nonexistent child-rearing habits, they might have ended up writing powerful verse about some other deep-seated biochemical urge—temperature regulation, say—but there would be no love sonnets in the reptile canon. The biological capacity for love is one way the brain prepares us for offspring who are born young and helpless, and who need tending to have even the slightest hope of survival. That tending comes in the form of social bonds: between parent and child, between the parents themselves, between the extended social family that helps raise the child. The glue that keeps those bonds strong is the feelings of pleasure and reward and satisfaction that our brains concoct for us when we enter into loving relationships.

When you think about love and attachment from this perspective, love starts to look like a kind of solution to an exceptionally difficult problem: getting organisms to take care of other organisms even if it's not in their direct best interest to do so. New parents will recognize this insight immediately; there are days (or more likely nights) when you look down at the screaming, defecating life-form on the changing table and you think, *Why am I doing this?* The neurocircuitry of love is evolution's way of persuading you to stick it out. Changes in your brain chemistry prod you to search out food when your belly is empty or safety when you're under attack. They also impel you to console your children and continue changing their diapers, despite the sleep deprivation and the temper tantrums.

The evolutionary biologist Donald Symons has an elegant explanation for how our emotions evolved: we have powerful feelings precisely because the goals our emotions are propelling us toward are difficult ones to achieve. The more difficult the objective, the more powerful the feeling. In the environments where our brains evolved, finding food and tending to children were extremely challenging tasks, yet vital to reproductive success. So evolution hit upon a way to encourage us, by creating reward circuits in the brain that made us relish both our offspring and our meals. Consuming oxygen is just as important to our reproductive success, of course, but oxygen is plentiful in our environment, so we don't have great feelings of warmth and contentment when we breathe. We need oxygen to live, but because it's not hard to get, we don't have an elaborate reward circuitry that propels us to search it out in the face of adversity.

Perhaps the best example of the evolutionary pressures of love on parent-child attachment connects back to our initial discussion of mindreading: the smiling instinct. All neurologically healthy children begin smiling sometime in their first few months of life, and most parents will testify that the appearance of those beaming faces marked a turning point in their relationship to their child. After weeks of oscillating between sleeping and crying, their child's eye and mouth muscles start to signal happiness, often when encountering a parent's face. Suddenly the child is capable of positive feedback—and it doesn't come a minute too soon. If you embarked on some monstrous genetic experiment and pushed back the onset of infant smiling six months, I suspect you'd see a proportionate rise in the number of children given up for adoption, or abandoned altogether. Those first smiling exchanges are some of evolution's most beautiful duets: a brain wired to produce a specific expression interacting with another brain wired to feel intense pleasure at the sight of that expression. They are the first unspoken phonemes in the language of love.

* * *

Smiles are only the beginning, of course. Many of the words in this book were typed with a constant sound track echoing through our home: our toddler son's laughter. On most days, it seems as though laughing is his primary activity. And the laughing effect is usually infectious—it's hard not to hear his delighted guffaws and not chuckle along with him. My wife and I do all sorts of rewarding things with our son, but few are as uniformly happy, as emotionally warming, as laughing together with him. Not surprisingly, our laughing is often triggered by tickling and rough-and-tumble play.

There's nothing new here—but something strange nonetheless. We take it for granted both that tickling causes laughter and that one person's laughter will easily "infect" other people within earshot. Even a child knows these things. But when you think about them from a distance, they seem like strange conventions. We can understand readily enough why natural selection would have implanted the fight-or-flight response in us, or endowed us with a sex drive. But the tendency to laugh when others laugh in our presence, or to laugh when someone strokes our belly with a feather—what's the evolutionary advantage of that? And yet a quick glance at the Nielsen ratings or the personal ads will tell you that laughter is one of the most satisfying and sought-after states available to us.

Understanding the roots of laughter requires that you undo your habitual assumptions about how "natural" it is. We're accustomed to thinking of laughter as a logical response to humor, but this connection is a misleading one. The closer we get to understanding what makes us laugh, the farther we get from humor. To appreciate the roots of laughter, you have to stop thinking about jokes.

There is a long, semi-illustrious history of scholarly investigation into the nature of humor, from Freud's *Jokes and Their Relationship to the Unconscious*, which may well be the least funny book about humor ever written, to a British research group that announced recently that it had determined the World's Funniest Joke. Despite the researchers' claim to have sampled a massive international audience in making this discovery, the winning joke revolved around none other than residents of New Jersey:

> A couple of New Jersey hunters are out in the woods when one of them falls to the ground. He doesn't seem to be breathing; his eyes are rolled back in his head. The other guy whips out his cell phone and calls the emergency services. He gasps to the operator: "My friend is dead! What can I do?"
>
> The operator, in a calm, soothing voice, says: "Just take it easy. I can help. First, let's make sure he's dead." There is silence, then a shot is heard. The guy's voice comes back on the line. He says: "OK, now what?"

As this joke illustrates, most assessments of humor's underlying structure gravitate to the notion of controlled incongruity: you're expecting x and you get y. For the hunting joke to work, it has to be readable on two levels, with two plausible ways to interpret the 911 operator's instructions—either the hunter checks his friend's pulse or he shoots him. The context sets you up to expect that he'll check his friend's pulse, so the—admittedly dark—humor arrives when he takes the wildly unlikely path. That incongruity has limits, of course: if the hunter chooses to do something utterly nonsensical—untie his shoelaces or climb a tree—the joke wouldn't be funny. As we'll see in the next chapter, it's no surprise that surprise plays a role here: the brain contains a number of subsystems that respond strongly to unexpected or novel developments.

A handful of studies in recent years have looked at brain activity while subjects were chuckling over a good joke—attempting, in a sense, to locate the neurological funny bone. Early evidence suggests that the frontal lobes are implicated in "getting" the joke, while the brain regions associated with motor control execute the physical response of laughter. One study analyzed patients with damage to the right frontal lobes, an integrative region of the brain where emotional, logical, and perceptual data converge. The brain-damaged patients had far more difficulty than control subjects in choosing the proper punch line to a series of jokes, usually opting for absurdist, slapstick-style endings rather than traditional ones. Humor may often come in coarse, lowest-common-denominator packages, but actually getting the joke draws upon our higher brain functions.

This is the kind of research that Robert Provine thought he'd be doing when he set out to study laughter: having people listen to jokes and other witticisms, and watching what happened. A professor of psychology and neuroscience at the University of Maryland, Provine is the author of the book *Laughter: A Scientific Investigation*—the culmination of a decade-long quest to determine why we laugh. Provine began by simply observing casual conversations and counting the times that people laughed while listening to another person speaking. But he quickly noticed a fundamental flaw in his assumptions about how laughter worked. "I started recording all these conversations," Provine says, "and the numbers I was getting—I didn't believe them when I saw them. The speakers were laughing more than the listeners. Every time that would happen, I would think, *Okay, I have to go back and start over again, because that can't be right.*"

Speakers, in turned out, were 46 percent more likely to laugh than listeners—and what they were laughing at, more often than not, wasn't exactly funny. Neither listeners nor speakers seemed to be laughing at traditional jokes. Provine and his team of grad students recorded the ostensible "punch lines" that triggered laughter in ordinary conversation. They found that only around 15 percent of the sentences that triggered laughter were humorous in any reasonable sense of the word. The big laugh lines included:

I'll see you guys later.
Put those cigarettes away.
I hope we all do well.
It was nice meeting you, too.
We can handle this.
I see your point.
I should do that, but I'm too lazy.
I try to lead a normal life.
I think I'm done.
I told you so!

The few studies of laughter that preceded this one had assumed that laughing and humor were linked inextricably, but Provine's early research suggested that the connection was only an occasional one. People certainly laughed at jokes, but that was only a small part of the story. "There's a dark side to laughter that we sometimes are too quick to overlook," he says. "The kids at Columbine were laughing as they walked through the school shooting their peers."

As his research progressed, Provine began to suspect that laughter was in fact about something else—not humor or gags or incongruity, but social interaction. He found support for this assumption in a study that had already been conducted, analyzing people's laughter patterns in social and solitary contexts. "You're thirty times more likely to laugh when you're with other people than you are when you're alone—if you don't count simulated social environments like laugh tracks on television," Provine says. "In fact, when you're alone, you're more likely to talk out loud to yourself than you are to laugh out loud. Much more." Think how rarely you'll laugh out loud at a funny passage in a book but how quick you'll be to make a friendly laugh when greeting an old acquaintance. Laughing is not an instinctive physical response to humor, the way a flinch responds to pain or a shiver to cold. It's an instinctive form of social bonding that humor is crafted to exploit.

Provine's lab at the Baltimore campus of the University of Maryland looks like the back room at a stereo-repair store—long tables cluttered with old equipment, tubes and wires everywhere. The walls are decorated with brightly colored posters of tangled neurons. (Add some Day-Glo typography and they might pass for signs promoting a Grateful Dead concert at the Fillmore.) A Tickle Me Elmo doll lies draped over a chair. Provine's old mentor, the late neuroembryologist Victor Hamburger, glowers down from a picture hung above a battered Silicon Graphics workstation. Hamburger's expression suggests a sense of concerned bafflement: "I trained you as a scientist, and here you are playing with dolls!"

While much of Provine's work draws on his training under Hamburger, exploring the neuromuscular control of laughter and its relationship to the human and chimp respiratory systems, the most immediate way to grasp his insights is to watch video footage of some of his more informal fieldwork, which basically consists of Provine and a cameraman prowling Baltimore's Inner Harbor, asking groups of people to laugh for the camera. The overall effect is more like that of a color story for the local news than serious research, but as Provine and I watch the tapes together in his lab, I find myself looking at the laughers with fresh eyes. Again and again, the same pattern repeats on the screen: Provine asks someone to laugh, and they demur, look puzzled for a second, say something like, "I can't just laugh." Then they turn to their friends or family, and the laughter rolls out of them as though it were as natural as breathing. The pattern stays the same even as the subjects change: a group of high school students on a field trip, a married couple, a pair of college freshmen.

At one point Provine—dressed in a plaid shirt and khakis, looking something like the comedian Robert Klein—stops two waste-disposal workers driving a golf cart loaded up with trash bags. When they fail to guffaw on cue, Provine asks them why they can't muster one up. "Because you're not funny," one of them says. They turn to each other and share a hearty laugh.

"See, you two just made each other laugh," Provine says.

"Yeah, well, we're coworkers," one of them replies.

The insistent focus on laughter patterns has a strange effect on me as Provine runs through the footage. By the time we get to the cluster of high school kids, I've stopped hearing their spoken words at all and hear just the rhythmic peals of laughter breaking out every ten seconds or so. Sonically, the laughter dominates the speech; you can barely hear the dialogue underneath the hysterics. If you were an alien encountering humans for the first time, you'd have to assume that the laughing served as the primary communication method, with the spoken words interspersed as an afterthought.

After one particularly loud outbreak, Provine turns to me and says, "Now, do you think they're all individually making a conscious decision to laugh?" He shakes his head dismissively. "Of course not. They're not aware of making a decision at all. In fact, we're often not aware that we're even laughing in the first place. We've vastly overrated our conscious control of laughter."

There is evidence that the physical mechanism of laughter itself is generated in the brain stem, the most ancient region of the nervous system, which is also responsible for fundamental life-or-death functions like breathing. Sufferers of amyotrophic lateral sclerosis—Lou Gehrig's disease—which targets the brain stem, often experience spontaneous bursts of uncontrollable laughter without feeling happiness or mirth. (They often undergo a comparable experience with crying as well.) Sometimes called the "reptilian brain" because its basic structure dates back to our reptile ancestors, the brain stem is devoted largely to our most primal, life-sustaining instincts, far removed from our complex, higher-brain skills at understanding humor. And yet somehow, in this primitive region of the brain, we apparently find the urge to laugh.

We're accustomed to thinking of common-but-unconscious instincts as being essential adaptations, like the startle reflex or the suckling of newborns. Why would we have a reflex for something as frivolous-seeming as laughter? Watching Provine's teenage laughers on-screen reminds me of an old Carl Sagan riff in which he describes "a species of primate" that likes to gather in packs of fifty or sixty, crammed together in a darkened cave, and hyperventilate in unison, to the point of almost passing out. The behavior is described in such a way as to make it sound exotic and somewhat foolish, like salmon swimming furiously upstream to their deaths or butterflies traveling thousands of miles to rendezvous once a year. The joke, of course, is that the primate is homo sapiens, and the group hyperventilation is our fondness for laughing together at comedy clubs or theaters, or with the virtual crowds of television laugh tracks.

I'm thinking about the Sagan riff when another burst of laughter arrives through the television's speakers, and without even realizing what I'm doing, I find myself laughing along with the kids on the screen. I can't help it—their laughter is contagious.

We may be the only species on the planet that laughs together in such large groups, but we are not alone in our appetite for laughter. Not surprisingly, our near relatives, the chimpanzees, are also avid laughers, although differences in their vocal apparatus cause the laughter to sound somewhat more like panting than it does the human variety. "The actual production seems a bit different between the two species because the chimpanzee's laughter is rapid and breathy, whereas ours is punctuated with glottal stops," says legendary chimp researcher Roger Fouts. "Also, the chimpanzee laughter occurs on the inhale and exhale, but ours is primarily done on our exhales. But other than these small differences in production, it seems to me to be just like ours in most respects."

Chimps don't do stand-up routines, of course, but they do share with humans a laughter-related obsession with humans, one that Provine believes is central to the roots of laughter itself: chimps love tickling. Back in his lab, Provine shows me video footage of a pair of young chimps named Josh and Lizzie playing with a human caretaker. It's a full-on ticklefest, with the chimps panting away hysterically each time their bellies are scratched. "That's chimpanzee laughter you're hearing," Provine says. It's close enough to human laughter that I find myself chuckling along with it as well.

Parents will testify that ticklefests are often the first elaborate play routine that they engage in with their children, and one of the most reliable laugh inducers. According to Fouts, who helped teach sign language to Washoe, perhaps the world's most famous chimpanzee, the practice is just as common, and perhaps more long-lived, among the chimps. "Tickling seems to be very important to chimpanzees because it continues throughout their lives," he says. "Even Washoe at the age of thirty-seven still enjoys tickling and being tickled by her adult family members." Among young chimpanzees that have been taught sign language, tickling is a frequent topic of conversation.

Like laughter, tickling is almost by definition a social activity. Like in the incongruity theory of humor, tickling relies on a certain element of surprise, which is why it's impossible to tickle yourself. Predictable touch doesn't elicit the laughter and squirming of tickling—it's unpredictable touch that does the trick. A number of tickle-related studies have shown convincingly that tickling exploits the sensorimotor system's awareness of the difference between self and other: if the system orders your hand to move toward your belly, it doesn't register surprise when the nerve endings on your belly report being stroked. But if the touch is being generated by another sensorimotor system, the belly-stroking will come as a surprise. The pleasant laughter as a result of tickling is the way the brain responds to that touch. In both human and chimpanzee societies, tickling usually first appears in parent-child interactions and has an essential role in creating those initial bonds. "The reason [tickling and laughter] are so important," Roger Fouts says, "is because they play a role in maintaining the affinitive bonds of friendship within the family and community."

A few years ago, the Pulitzer Prize–winning scientist Jared Diamond wrote a short book with the provocative title *Why Sex Is Fun.* Some of the research into laughter suggests an evolutionary answer to the question of why tickling is fun: it encourages us to play well with others. Young children are so receptive to the rough-and-tumble play of tickle that even pretend tickling will often send them into wails of laughter. (Fouts reports that the threat of tickle has a similar effect on his chimps.) In his book, Provine suggests that "feigned tickle" can be thought of as the Original Joke, the first deliberate behavior in a child's life designed to exploit the tickling/laughter circuit. Our comedy clubs and sitcoms are culturally enhanced versions of those original playful childhood exchanges. Where we once laughed at the surprise touch of a parent or sibling, we now laugh at the surprise twist of a punch line. Along with the suckling and smiling instincts, the laughter of tickle evolved as a way of cementing the bond between parents and children, creating an impulse that eventually carried over into the social lives of adults.

Bowling Green University professor Jaak Panksepp, one of my mentors on the neuroscience of emotions, has gone so far as to suggest a dedicated "play" circuitry in the brain, equivalent to the more extensively studied fear or love circuits. Panksepp has studied the role of rough-and-tumble play in cementing social connections between juvenile rats. The play instinct, he has discovered, is not easily suppressed: rats that have been denied the opportunity to engage in rough-and-tumble play—which has a distinct choreography among young rats, along with a chirping vocalization that may be the rat equivalent of laughter—will nonetheless immediately engage in play behavior given the chance. Panksepp compares this inclination to a bird's instinct for flying. "Probably the most powerful positive emotion of all—once your tummy is full and you don't have bodily needs—is vigorous social engagement among the young," Panksepp says. "The largest amount of human laughter seems to occur in the midst of early childhood—rough-and-tumble play, chasing, all the stuff they love."

Playing is what young mammals do, and in humans and chimpanzees, laughter is the way the brain expresses the pleasure of that play. "Since laughter seems to be ritualized panting, basically what you do in laughing is replicate the sound of rough-and-tumble play," Robert Provine says. "And you know, that's where I think it came from. Tickle is an important part of our primate heritage. Touching and being touched is an important part of what it means to be a mammal. I mean, this is why we're not lizards!"

There is much that we don't yet know about the neurological underpinnings of laughter. We do not yet know precisely why laughing feels so good, though one recent study detected evidence that laughing triggered activity in the nucleus accumbens, the same region implicated in love circuitry. Panksepp has performed studies that suggest drugs that block the effects of opiates suppress the play instinct in rats, which implies that the brain's endorphin system may be involved in the pleasure of laughter. Anecdotal evidence along with some clinical studies do suggest that laughing makes you healthier by suppressing stress hormones and elevating S-IgA immune system antibodies. If you think of laughter as a form of human behavior that is basically synonymous with the detection of humor, the laughing-makes-you-healthier premise seems bizarre. Why would natural selection make our immune system respond to jokes? Provine's approach helps resolve the mystery: our bodies aren't responding to wisecracks and punch lines, they're responding to social connection.

This approach to laughter provides a fascinating bridge between evolutionary and Freudian psychologies. In both models the past weighs heavily on the present-tense brain. In the psychoanalytic model, it's the anxieties and trace memories of childhood haunting the adult psyche. In the Darwinian model, it's the ancestral environment where our brains evolved that does the haunting: we live in cities and suburbs now, but our brains are filled with tools optimized for the savannahs of Africa. In both models, our past complicates our present reality, because the drives and appetites of the past aren't always in sync with those of the present. In both we are haunted by our origins: the childhood of the species in the Darwinian framework and the childhood of the individual in the Freudian.

Understanding the roots of laughter requires a kind of hybrid of the Darwinian and Freudian models. We laugh primarily because laughter is a crucial component of the emotional glue that connects parent and child during the most vulnerable years of development. Children who laugh and roughhouse and tickle with their guardians create powerful bonds of affection with those grown-ups, and the bonds help them survive. But natural selection is notoriously conservative with its designs: when you build a mechanism for bonding into the child's brain, the accompanying impulses don't necessarily disappear in adulthood or when children aren't around. So the difficulties of child-rearing created the capacity for—and the deep pleasure of—laughing, and once that capacity was installed, we came upon other applications for it. So when we laugh at the Chaplin film, we have childhood to thank for it. Not our individual childhood in the Freudian sense, but *childhood itself* and its unique challenges.

The idea that laughter evolved first to cement social connections and was only later hijacked by the stand-up comics is a particularly crucial insight in a world of ever-proliferating communication channels. Not long ago, I attended a small retreat on the design of communications software that put twenty-odd people in a room together to discuss various ideas face-to-face, while simultaneously letting them converse in a special electronic chat room restricted solely to the people attending the retreat. The chat was projected onto a flat screen visible to everyone in the room, and people typed their comments via their laptops.

The chat turned out to be a mix of follow-up observations and references to related reading on the Web, as well as the usual snarkiness that you'd expect. From one angle, it was a pretty intoxicating mix—the carrying on of two simultaneous group conversations with the same group. I felt like we were pulling down a lot of data: the real-world conversations grounded things, and the chat let the room riff. It was also a little intoxicating in the dizzying sense. Cognitive scientists have long known that our attention buffers max out when following two "verbal" conversations at once, and this experiment made me wonder if the carrying capacity is any different if one conversation is spoken and one is text.

But the most interesting side effect of this discussion was that the arrangement sucked all the jokes out of the room and into the chat. If someone had a funny throwaway remark to make, they'd simply toss it into the chat log. You'd see people smile to themselves as the joke scrolled across the screen, but they wouldn't laugh out loud. I mentioned this point near the end, as we were discussing the format, and someone said that having the jokes in the virtual world improved matters: the jokes were there for all to see, but they didn't interrupt the flow of the conversation. That observation was true enough, but only if you think the point of jokes is humor rather than laughter. If laughter is primarily a form of social bonding, then depriving the room of laughter will have a dramatic effect on its general tone. At the end of the session, when the moderator asked us to close our laptops and reflect on the day a bit, the space quickly reverberated with group laughter, which completely transformed the social climate. The number of jokes per minute probably declined, but the room felt far more collegial and cohesive. And that is because in the initial arrangement, with the humor stashed away on the digital screens, our brains had been deprived of the reward chemicals triggered by laughter. Jokes on their own simply weren't enough.

The lesson here is twofold. First, certain social settings—particularly those that involve virtual communication—may artificially dampen laughter that would otherwise be generated in a face-to-face encounter. Second, social interaction without laughter produces modified brain chemistry, which affects both your background impression of the exchange—its emotional color—and the resulting trace memories the exchange leaves in your head. Putting smiley faces into email to supplement the lack of verbal intonation helps convey when you're trying to be funny, but because the recipient of your message is still alone when reading it, she won't be likely to laugh out loud, and that suppressed laughter will make a difference. The memory will be happier—and consequently stronger—if she laughs.

As the brain science of social connection becomes more widely appreciated, our communications tools will be judged increasingly with this yardstick. Attention deficit disorder is conventionally described as the classic ailment of our multitasking age, but when you look at most electronic communication through the lens of neuroscience, it's hard not to think that autism might be a more appropriate "poster condition" for the digital society. (The cultural critic Harvey Blume made this argument nearly a decade ago.) When we interact with other humans via communication channels that are stripped of facial expressions and gestures and laughter, we are unwittingly simulating the blank emotional radar of the mindblind.

But for most people, I suspect, the neuroscience of personal connection will have more intimate revelations as we come to understand and recognize the chemicals that trigger these powerful feelings. Not just because it's intellectually interesting to know that your feelings of attachment are partially instigated by oxytocin, but also because the chemistry's effects go beyond the primary emotion itself—altering your memory, your immediate attention, your evaluation of people and environments. You can think of these areas as being like the side effects associated with pharmaceutical drugs, though this shouldn't imply that the effects result from poor design. (We'll turn to the whole idea of decoding the brain's chemical side effects in the next chapter.) When you begin to explore these peripheral effects on the mind, you're not just memorizing drug names; you're learning to recognize symptoms. Odds are, you've detected these symptoms before without knowing a thing about brain science, but you may have attributed some of the subtler symptoms to other causes, or found them hard to explain.

This was the story of our little family on 9/11. About a year after our son was born, I was talking with Sue Carter about her investigations into oxytocin, and I told her about my wife's strange sense of calm in the midst of all that chaos. It struck a chord immediately. "I'm very interested in breast-feeding as a protective mechanism, because of my whole experience nursing my own children," she said. Carter had in fact completed a number of studies on the topic, and the results explained precisely what was happening to my wife on that insane day.

"We compared the effects of stress on lactating and nonlactating women. With the lactacting women we know they have more oxytocin, and we know they manage stress better," she explained. Since Carter's studies were first published, additional research has convincingly demonstrated that oxytocin is what scientists call a "downregulator" of the body's HPA system, the circuitry that creates the bleak, gut-tightening feeling you experience when you get the news that the promotion didn't come through—or when CNN reports that another plane is missing. People under the influence of oxytocin don't have the same stress responses that others do; bad news rolls off them more readily.

That's the tending instinct for you. You can fight your way out of stress by destroying your enemies, or you can reduce it by reaching out to loved ones. As far as brain chemistry goes, there are two strategies available: you can load up on adrenaline and fight-or-flight, or you can cool down with oxytocin and tend-and-befriend. I'd unconsciously opted for one strategy, my wife the other. Where our brain chemistry was concerned, we were on two different rides.

There's a risk for easy reductionism in this domain. Before I leave her office, Shelley Taylor cautions, "A lot of people say, 'Oxytocin is the cuddly hormone,' or, 'Oxytocin's the love hormone.' Oxytocin is much more evasive than that, and it doesn't have one-to-one correspondence with psychological states. It's real risky trying to map these molecules onto specific states."

"For instance," she says, leaning forward in her chair for emphasis, "older women who are living with husbands and finding those husbands to be nonsupportive have chronically higher levels of oxytocin. Now it's not clear the direction of causality. But a tentative conclusion that I would make is that when social support needs are not being met, oxytocin levels go up as a signal to seek out social contact. And then once found, it may be restored to normal levels. So oxytocin isn't the 'feel good' hormone. It may be the 'feel crummy' hormone that leads you to take steps to feel better."

Certainly the passions of love are a mix of brain chemicals, not just oxytocin. Some scientists believe that oxytocin works in tandem with the body's natural opiates, with oxytocin triggering the drive for social attachment and the opioids supplying the "warm fuzzy" feeling that you get in the company of loved ones. Jaak Panksepp believes that one effect of oxytocin on the body's "natural high" is to reduce the tolerance that plays such a devastating role in drug addiction. Just as addicts develop a tolerance to heroin that causes them to take ever-larger doses to reach the same high, the brain develops an identical tolerance to naturally occurring opiates. But in tests with animal subjects, oxytocin injections dramatically reduce tolerance to opiates. In other words, it's possible oxytocin does not create the visceral pleasure of love and attachment, but it does enable that pleasure to last longer than it normally would.

So the phrase "addicted to love" turns out to be more than mere poetry. Think of the similarity between brain scans of mothers listening to their babies crying, lovers gazing at photos of their partners, and drug users experiencing a cocaine rush. In each of these three situations, the external reality is quite different, but the *internal* chemistry is surprisingly alike. Drugs like heroin and cocaine do their damage because they tap directly into the brain chemistry that regulates the bonds of love. When people become addicted to drugs, one of the most common reactions expressed by close friends is bewilderment at the addict's ability to turn his back on the affiliations of family and friendship. Not knowing firsthand the tremendous force of addiction, many of us find it monstrous that someone should trade a child's love for the prick of a needle. But that needle contains the very drug that helps make a child's love appealing. We understand intuitively why someone might sacrifice his life for a child. When drug addicts make comparable sacrifices, it seems positively inhuman. And yet, neurochemically, those sacrifices are laid at the same altar.

Knowing something about the chemistry of love brings us closer to the gruesome worldview of addicts, and in doing so it can't help humanizing people in the throes of addiction. (Certainly it should make them seem less like criminals.) But for me, the longest-decay idea in the chemistry of love is the connection it opens up to newborn children. In the first few months after our son was born—as the smoke literally cleared after 9/11—my wife and I often found ourselves wondering whether it was fair to say that our son *loved* us the way we had grown to love him (despite the sleepless nights and endless diaper changing). We could tell that our presence—particularly my wife's—had a noticeably positive effect on him, pulling him down from heights of fierce howling to contented cooing in a matter of seconds. He smiled at us more than at anyone else, and as the months passed, he would occasionally appear frightened in the company of strangers. There was attachment there, to be sure, but it seemed a stretch to think of it as love. It was some other newborn emotion, we thought, perhaps as different from our grown-up feelings of love as it was from sadness or pain.

Our second boy was born during the final stages of writing this book, and in the two years that had passed, I had learned enough about the neurochemistry of love to formulate a reasonable answer to the question my wife and I had wrestled with. I no longer see that early attachment as a distinctly newborn emotion, separate from our adult feeling. The experience—the qualia—of grown-up love is shaped by a thousand memories flashing through your head as emotion flows through you: memories of past loves, of romantic poetry, of Audrey Hepburn movies, and, most of all, of memories of the person who triggers the feeling in you. Newborn children haven't lived long enough to assemble all their memories, and they don't have a system developed enough to record or play back that remembered complexity. Grown-up love, though, is also a chemical feeling, one that has effects on our memory systems, but also one that possesses a life of its own. We don't know the exact ingredients of the cocktail, and no doubt the cocktail differs from person to person in the exact ratios. But some mix of oxytocin and endorphins is clearly pivotal to the feeling of love. They conjure up that sense of warmth and contentedness, the sense of being where you're supposed to be. That feeling is not the whole story of love, of course, but it is a dominant thread.

I believe it's this chemistry that we share with our children, even children in their first days of life. When our younger son switches from tantrums to giggles at the sight of his mother entering a room, he's doing so because the sight of her face has released a host of chemicals in his head—the same chemicals flooding through his mother's brain as she gazes back at her child. The infants don't have words for the feeling, and for them it isn't accompanied by the rich tapestry of memories invoked by grown-up attachment. But some essential part of the feeling is mirrored in those two brains. It's nice to think that each of us has unique ways of feeling love, but there are times when the shared experience is even more moving. Parents and their newborn children don't yet have a language in common, and they have almost no past together to remember. But they are capable of sharing nonetheless, precisely because the chemistry of love has a common design. At some point in your first days of life, your brain began sending signals to you saying, "You're safe with this person; keep close to her." Decades later, you're still getting the same message.

5

The Hormones Talking

"In one way or another, all our experiences are chemically conditioned, and if we imagine that some of them are purely 'spiritual,' purely 'intellectual,' purely 'aesthetic,' it is merely because we have never troubled to investigate the internal chemical environment at the moment of their occurrence."
—ALDOUS HUXLEY

No fact unearthed by modern neurochemistry has circulated as widely as the existence of naturally occurring pleasure drugs in the brain. Brain researchers had long suspected that the family of painkilling drugs derived from the opium poppy—heroin, morphine, codeine—targeted a dedicated site in the brain, but it wasn't until the early 1970s that a handful of researchers working in separate labs discovered the receptor: a synaptic lock contoured precisely to fit the opiate keys. This was one of those discoveries whose existence suggests a further discovery. Though the narcotic allure of the poppy has been part of human experience since at least the dawn of agriculture, it seemed unlikely that the brain would possess a receptor for a chemical found in a plant that grew in only a few scattered locations around the globe. The existence of the receptor suggested that the brain produced its own endogenous opiate, and sure enough within a few years scientists had discovered two of them: the enkephalins and the endorphins—meaning "in the head" and "morphine within," respectively. Newspapers, magazines, and talk radio were flooded with excitement over the brain's "natural high." And the surge of interest in fitness and jogging that began twenty-five years ago owes its existence in part to the discovery that these powerful chemicals were released during strenuous activity. People don't get in shape simply because it's in their long-term interests. They get in shape because working out makes them feel good, and their brains remember that feeling.

The endogenous opioids were not alone. Receptors have now been discovered for the active ingredients in marijuana, nicotine, and the underground psychedelic DMT. Even chocolate turns out to possess a naturally occurring chemical, phenylethylamine, and it may activate some of the same receptors that help create the marijuana high.

In a funny way, the existence of these receptors plays against the propaganda of both the drug subculture *and* the antidrug movement. There's a classic argument that you may have encountered before—in conversation or in print—that usually comes from someone evangelizing a psychedelic like DMT or "magic mushrooms." The fact that the brain contains a DMT receptor is seen by them as evidence of a higher purpose, somehow making the drug more portentous and revelatory than it would otherwise be. But the argument is a tautology. Of course these drugs have targeted receptors. That's what makes them drugs. They're "psychoactive" because they can pick receptor locks in the brain—otherwise, they wouldn't "activate" the "psyche." There are millions of plant species on earth that do not contain molecules that mimic the brain's endogenous chemistry. The fact that a small number have evolved overlapping chemistries—and have thus been cultivated extensively by the people who discovered them—can be explained by everyday statistics: given enough plants, and enough people experimenting with the plants, some will contain chemicals that pass for our own endogenous drugs. If people ingesting these plants enjoy the experience, they'll spread the genes of the plants by cultivating them and growing crops. The rest is just agriculture.

By the same token, the very idea of endogenous drugs complicates the easy rhetoric of the "drug war." The legendary phrase from the Reagan years—"This is your brain on drugs"—is ultimately misleading. Your brain is nothing but drugs—or put another way, it would be nothing *without* drugs. Certainly there is a distinction to be made between those that are endogenous and exogenous, between natural and artificial, but the fundamental truth is that artificial drugs work because your brain mistakes them for natural ones. Right now, as you read these words, you are under the influence of chemicals that are, molecularly speaking, almost indistinguishable from drugs that could get you arrested if you consumed them openly in a public place.

This is not necessarily a prolegalization argument. Our brains are vastly better at regulating the release and reuptake of endogenous drugs than of exogenous ones. Indeed, one of the ways that recreational drugs achieve their potency is by short-circuiting the brain's normal maintenance work. No one overdoses on endorphins, but thousands die every year of heroin overdoses. Regular abuse of these substances can cause long-term neurological damage, as brain scans of chronic speed or cocaine users have amply demonstrated. One study scanned dopamine receptors in the brains of alcoholics, overeaters, and cocaine addicts. In scans of healthy brains, the dopamine-rich areas showed up as two symmetrical bright red blotches, fading out to green at their peripheries. (Red signals the most active areas.) In the brains of alcoholics and overeaters, the red spots were a fraction of the size of those in the normal brain, meaning that they received a reduced dose of the brain's natural supply of dopamine. In the brains of cocaine addicts, the red spots were nonexistent.

The damage inflicted by long-term drug use is significant enough that it's hard to argue against society trying to prevent people from abusing serious drugs like cocaine and amphetamines. But we should also not be so quick to imagine the pleasures of drug use as some alien, unnatural experience, far outside the boundaries of the "straight" world. Beneath all the drug war rhetoric there is this sobering—or is it intoxicating?—thought. If you could take fMRI snapshots of your brain at the happiest moments of your life, the images would probably look remarkably similar to brain scans of people doing heroin or cocaine for the first time.

So begin with this basic premise: you are on drugs. With every shifting mood, every twitch of anxiety, every lovelorn glance, you are experiencing the release of dedicated chemicals in your brain that control your emotions, chemicals fundamentally the same as the ones you might otherwise find in a dime bag or a coke spoon. The potency of most recreational drugs is largely a matter of quantity, not quality: your brain just doesn't have an internal delivery mechanism that can flood your synapses with as many molecules as ingested drugs can—for good reason.

In a few isolated situations, we already accept the premise that our behavior is shaped by endogenous drugs, as in the sexist phrase you'll sometimes hear said of menstruating or pregnant women: "That's just the hormones talking." This is one of those pieces of pop-sci brain trivia that confuse as much as they enlighten. It's true enough that women's moods and perceptions are altered by chemicals released when they menstruate and during pregnancy. But dismissing those feelings as "just hormones" creates two misleading impressions. First, it implies some kind of opposition between a woman's "normal," or "true," personality and the effects triggered by the release of hormones. Yet as we have seen again and again, without hormones (and other brain chemicals, like neurotransmitters) none of us would have personalities in the first place. When your brain isn't markedly guided by estrogen or oxytocin, it's still under the influence of dopamine, serotonin, and all the rest. At any given time, your background moods and foreground emotions are a measure of the various chemicals swirling around in your head. To a certain extent, it's always the hormones talking.

This gets to the second misconception at the root of the sexist remark. The idea that women go through distinct phases during which they are "hormonal" creates a false opposition between a chemically manipulated female brain and a sober, unalterable male brain, free from the irrational influence of hormones. Testosterone alters male judgment and behavior as dramatically as estrogen does, and while you'll sometimes hear people described as being "high testosterone," you'll rarely hear someone dismiss an aggressive male CEO's performance as being "just the hormones talking." The roots of the "hormonal" dismissal may dwell in the fact that variations in women's neurochemical makeup show themselves in more regular cycles than men's do. Such regularity creates a pattern that becomes detectable over time: certain days each month, your mood changes. Thus, on those days, you think of yourself as "hormonal," while the rest of the time you're just you.

The goal of recreational neuroscience shouldn't be to do away altogether with the idea of hormonal influence. The goal should be to understand it more precisely. The question shouldn't be "Are the hormones talking?" It should be "*Which* hormones are talking, and what are they saying?"

The first step to answering these questions is learning to recognize the release of specific chemicals in your brain. You probably know some examples of this already: the surge of adrenaline after a sudden fright; the heady social confidence prompted by serotonin after being elected class president. The more you pay attention to these chemical releases, the better you get at detecting them. In the language of music appreciation, you develop an ear.

The audio analogy is a helpful one, in fact. Many people who listen to pop music can't easily separate out the bass line the way they can the vocals or the drums or the lead guitar. The bass just blends into the overall mix: listeners *feel* it on some level, and do notice something drop out of the soundscape when it is removed using an equalizer. But they can't hear the bass notes themselves as distinct entities. This is where ear training comes in. In particular, when someone hears the isolated bass line of a song, and then hears the rest of the mix superimposed, the bass acquires a new clarity. And the amazing thing is that once you've developed an ear for the bass, you can't *not* hear it. At first, it's for all practical purposes silent. And then you can't shut it off.

Brain chemistry is like that bass line: once you've learned to recognize certain chemicals, those chemicals start to pop out at you, as clear as a headache or a dizzy spell. That knowledge certainly makes you feel like a more discriminating user of your brain, but there's a bit of the neuromap fallacy lurking here as well. Does it change anything essential to know the names for "cortisol" and "oxytocin"? Is that any more helpful than knowing the location of the craving center?

If understanding brain chemistry were simply a matter of memorizing the nomenclature, it would be little more than cocktail party banter. But learning about your internal brain chemistry is more than just memorizing terms. Most important, it involves learning and recognizing the side effects and subtle properties of your body's drugs. You can't necessarily erase those side effects just by understanding them, but you can put them in context, and anticipate the ways in which they will likely alter your judgment.

Consider these two scenarios. Imagine that you knowingly take a dose of hallucinogenic mushrooms. After an hour or so, you begin to feel a growing sense of sensory confusion; colors and sounds swirl together; you feel sudden surges of insight and equally sudden bouts of fear; elaborate dancing patterns fill your field of vision if you close your eyes. You may even hallucinate entire creatures interacting with you in various implausible ways.

Now imagine that you *unknowingly* take that same dose, and suddenly your world is transformed for no apparent reason. The hallucinations and dramatic mood swings descend on your brain seemingly out of nowhere.

In each case, your brain chemistry is being altered by the exact same drug, and yet the two experiences would most likely be dramatically different. The first might well be exhilarating and illuminating (though of course it could also break down into "bad trip" paranoia). The second scenario, however, would almost certainly feel like a decidedly unpleasant form of madness.

The difference between the two situations is simply this: the drug doesn't change, but your *awareness* of the drug and its effects does. Knowing that mushrooms are capable of transforming an Oriental rug into a writhing nest of snakes makes it far easier to enjoy the snakes—and recognize their illusory nature. You can't coerce your brain into stopping the hallucination, but you can comfort yourself with the knowledge that the hallucination is a normal effect of the drug you've taken. And this comfort will doubtless change your behavior in profound ways: instead of racing out of your house screaming "The snakes! The snakes!" or checking yourself into a mental institution, you sit on the couch and giggle. Understanding the full range of a drug's effects changes the experience of taking it.

The same is true of endogenous drugs. Knowing the stress-management effects of oxytocin would have helped both my wife and me understand her strangely distanced state on the morning of 9/11. As I learned more about the amygdala's capacity for triggering adrenaline surges when reminded of traumatic events, I found it easier to deal with the conditioned fear response I felt every time the wind started howling outside our apartment windows. I'd feel the anxiety start to well up inside me, and I'd think, *It's just the hormones talking.* The feeling wouldn't exactly go away, but its effects would be less paralyzing.

Learning to recognize the range of effects unleashed by our naturally occurring drugs—"listening" to them, in Peter Kramer's famous phrasing—can also bring new psychological categories into focus. Kramer himself described such a category in *Listening to Prozac:* rejection sensitivity. No one likes being rejected, of course, but people with acute rejection sensitivity have unusually strong reactions to disappointing news or personal slights—sometimes they'll go out of their way to avoid situations in which rejection is a possibility, hindering their ability to take risks that might on average make their lives better. The standard psychological manuals had no category for this condition until a new family of drugs came along and made it visible.

It turns out that the serotonin system plays a key role in modulating rejection sensitivity. Because Prozac increases the amount of serotonin available to the brain, rejection-sensitive people found that their original vulnerabilities faded away under the influence of the drug. They didn't feel euphoric or recklessly giddy; they didn't lose their sense of perspective. The change was more subtle than that: small slights and minor social disappointments rolled off them more easily. They didn't dwell on bad news as much, which left them feeling more confident, more inclined to risk rejection in the future, when appropriate.

I suspect that for many people who read *Listening to Prozac*—or who went on Prozac themselves—the idea of rejection sensitivity was one of those attributes that they had experienced in a background way but had never before really put their finger on. You're generally happy, feeling motivated and engaged with the world, but the one thing that's bothering you is that you're just a little too fragile when something goes mildly wrong, particularly in social situations. You're not depressed, or manic, or obsessive-compulsive, but you're easily deflated by events that shouldn't really deflate you all that much. Because that fragility didn't have a name, most of us didn't think about it all that much, at least the way we thought about other pop-psych categories: extroverts and introverts, right brain and left brain people, manic-depressives and schizophrenics. What illuminated the category of rejection sensitivity was our newfound ability to target serotonin reuptake channels, and thus manipulate serotonin levels with an unprecedented precision. When drugs make more serotonin available in the synaptic channels of people's brains, and keep everything else pretty much the same, people's rejection sensitivity diminishes. The change makes it easier to detect that particular tendency of mind. It's like that bass line you've developed an ear for, something you don't even have to listen for anymore. It just pops out at you.

Recognizing the peripheral effects of the brain's emotional system doesn't always involve tracking individual chemicals. The major neurotransmitters are themselves each involved in many forms of brain activity, while our felt emotions are the sum total of dozens of physiological and chemical changes in our body. The language of emotion is filled with references to the body: our skin crawls, our hearts race. These are not only metaphors: the chemicals released during emotional states trigger specific events throughout the body—so much so that William James famously argued that emotions were nothing more than the aggregate of those bodily changes. You don't feel fear and then feel your heart start to race—fear, James proposed, *is* your heart racing.

But while emotional systems can't be reduced to single "magic bullet" chemicals, they nevertheless produce reliable peripheral effects, and learning to recognize those effects can make it easier to inhabit your own head. Several years ago, Antonio Damasio led a team of researchers in a study that performed PET scans on people's brains as they recalled intense emotional experiences from their past. The subjects were asked to relive as vividly as possible events that involved happiness, sadness, fear, or anger. When the subjects felt the emotion consume them again, they signaled to the researchers. Damasio and his team were trying to determine the regions of the brain responsible for creating the feeling itself—in other words, which parts reported on the changes in the body's physiological state brought about by the emotion. And indeed they found dedicated areas that lit up at the exact moment the emotion was felt, and that each emotion created a precise neuromap that could be readily distinguished from the others.

But the researchers also stumbled across another observation that hadn't been central to their experiment. Sadness was marked by decreased activity in the prefrontal cortices, while happiness triggered an increase in such activity. Prefrontal cortical activity is a strong predictor of idea generation and overall liveliness of thought. When you're thinking on your feet, when you're full of ideas, your frontal lobes are firing on all cylinders. What Damasio found was that happiness elevated those firing rates, while sadness dampened them. In other words, one of the side effects of the way the brain creates the feeling of sadness is a reduction in the overall number of thoughts that the mind produces.

When I first read about Damasio's study, this finding struck me immediately as liberating. I thought of all the times over the years when I'd been feeling blue for some reason, and while wallowing in my mood, I'd note that I hadn't had an interesting idea in a disturbingly long time. My sadness would quickly deepen into a gloomy self-doubt: not only was I blue, but I was also becoming stupid! It was hard enough being sad, but now I had to deal with being dim-witted as well. Contrast that downward slide with my usual response to coming down with a head cold. Being sick makes me feel dense as well, but it rarely troubles me because I've always assumed that my brain was busy marshaling forces to ward off an invading virus and didn't have time for generating ideas. Being stupid was part of being sick, I assumed, and so I'd just hunker down in front of *Wheel of Fortune* and ride it out.

Damasio's study helped me see a pattern in my own psyche as a side effect of the chemistry of sadness, a side effect that wore off as soon as the emotion did. Once my blue mood passed, my prefrontal cortices would jump back to life, and I'd be back to my normal idea-generating clip. Since learning this, I've experienced the mental sluggishness of feeling sad without the downward cycle of self-doubt. Instead of wondering whether I've lost whatever mental agility I once had, I just wait it out. I have only anecdotal evidence for this conclusion, but I feel like my bouts of sadness have grown shorter with this new awareness, because the cycle of self-doubt has been eliminated.

Perhaps the most important insight to come out of the growing understanding of our brain's chemistry is what researchers call "mood congruity." Because the brain is an associative network, and because our memories record not just specific details of events but also our feelings about them, when the brain is under the influence of one emotion, it habitually makes connections to past events that triggered the same emotional response. When you're experiencing stress, your brain is more likely to recall stressful memories from your past than it is upbeat ones. When something frightens you, your mind is more likely to become filled with thoughts of other, apparently unrelated threats than it is examples of feeling safe. This is the essence of mood congruity: your memory system tends to serve up recollections of past events that are themselves congruous with your current mood.

The brain's architecture is designed in such a way that it does not play emotional devil's advocate. When you're filled with happiness and good cheer, your memory system doesn't remind you of that upcoming tax filing or your fear of getting fired from your job. If you're happy, you're more likely to think of the vacation that's just around the corner, or how much money you made on that stock trade last week. The brain doesn't do checks and balances. If the glass is looking half full, it pours a little more water in just to exaggerate the effect.

These self-perpetuating cycles partially explain why being happy is so much fun, and why depression can be so devastating. Severely depressed people have to be reminded actively that there are good things in their life; happy memories just don't pop into their minds the way they do in the minds of nondepressed people. This can be the case even if the stimulus that began the depressive cycle was fleeting, or altogether illusory.

Several years ago, doctors at the Salpêltrière Hospital in Paris were experimenting with a revolutionary new treatment for Parkinson's disease that involved implanting an electrode in a section of the brain stem that plays an important role in motor control. Most Parkinson's sufferers experience a decreasing ability to initiate movements, along with tremors and shakes; stimulating certain brain stem regions creates a marked decrease in these symptoms.

With one patient, however, the doctors accidentally stimulated an area that initiates the physical posture of great sadness. Within seconds of receiving the electric current, the patient slumped in her chair, a morose expression spreading across her face. Soon her eyes filled with tears, and her verbal report to the doctors was suddenly something straight out of Dostoyevsky's *Notes from the Underground*: "I'm fed up with life, I've had enough. . . . Everything is useless." When the doctors switched off the current, her despair disappeared almost instantly: she smiled and professed bafflement about why the world had suddenly seemed so bleak.

This incident may be the ultimate example of the power of emotional self-perpetuation. The Parkinson's patient had launched herself into abject misery without any *external* cause. The trigger was not experiencing something sad, or thinking sad thoughts. It was an electrical stimulation that triggered only the physical posture of sadness, and that bodily transformation was enough to fill her brain with powerfully miserable images. The associative matrix in her head was filled with memories of her body taking on this precise configuration—slumped shoulders, tears welling up—in response to genuinely saddening stimuli. And so when the electrodes recreated the configuration marionette-style, those memories obligingly flooded her worldview, and within seconds she had lost even the will to live.

We are all a little closer to that Parkinson's patient than we might like to believe. The feeling that everything just seems to be lining up for us after we get a piece of welcome news, or the sense that death and illness are everywhere after we attend the funeral of a good friend—these phenomenon are both usually illusions, conjured up by the brain's knack for association. Our emotional state skews our sense of perspective by seeking out memories that match our current mind-set instead of a balanced, representative sample. It's what the military calls "incestuous amplification." Your memory system is like the obliging colonel who squashes all bad news to tell the commander exactly what he wants to hear. As in this military analogy, if you know that the self-perpetuating cycle is under way, you can route around it: either by actively seeking out contrasting information or by simply taking your overall worldview with a grain of salt. Life really *isn't* as great as it feels right now. That's just the hormones talking.

Emotions do not merely mark certain memories as being more important than others. They also affect which details get recorded. Several years ago, the Harvard psychologist Kevin Ochsner conducted a study in which college students were exposed to a series of images, some of them positive in nature (a smiling child), some of them negative (a disfigured face), and some neutral. As you might expect, when the students' memories of these images were tested several days later, the images that had triggered strong emotional responses were recalled more readily. Positive images were as familiar as negative ones, while the neutral images had largely faded from memory. The brain's emotional underlining had worked as expected.

But Ochsner discovered a fascinating distinction when he probed his subjects for details about what exactly they remembered. For the positive images, the students recalled a general impression of the scene, along with a trace memory of the pleasant emotional response it triggered in them. But in recollecting the negative images, they remembered far more details. With happy images, they took in the gestalt of the scene presented, but they recorded the specifics of the disturbing images as though they were forensic pathologists examining a crime scene. Both types of images were encoded in memory, but the encoding process itself seemed to operate under different rules depending on whether the memory was positive or negative. This is another way of approaching the "flashbulb memory" phenomenon we saw in our discussion of the amygdala. When we're experiencing something disturbing, our brain takes in as many details as possible, just in case one of those details turns out to be relevant to a future threat.

This may be one reason why negative memories are more likely to haunt us than positive ones. The brain is an associative network, with memories represented by clusters of neurons firing in sync with one another. Sometimes overlapping clusters get triggered along with the original group—that's your brain expressing a connection between two associated memories: the sound of an old Patti Smith song reminding you of the college dorm where you first heard it; the bright blue sky of a clear autumn day reminding you of planes crashing into skyscrapers. If negative memories are built out of multiple details, each its own resonating cluster, then that equals, in a sense, more hooks in the brain to pull you back to the original disturbing event. Bad memories simply give you more details to remember.

These emotional side effects deserve to be more widely understood. Recreational drugs have well-known effects on our memory systems. LSD can create flashbacks that invoke details of a past trip, even one that occurred months before, with sometimes overwhelming intensity; marijuana impedes short-term memory; caffeine heightens our faculties for recall (at least during the first cup). These properties are widely recognized, despite the fact that a relatively small percentage of the population actively uses marijuana or LSD. But we all use the drugs that create positive and negative emotional responses in our brains: we all feel fear, and remorse, and good cheer. Those neurochemical reactions in our heads have predictable effects: they make stronger memories; they remind us of like-minded memories; they record more or less detail depending on their emotional valence. You've probably experienced the effects of mood congruity several times today; odds are, you've never experienced an acid flashback in your life. But which effect is a household term?

* * *

Our brains don't just underline positive and negative memories. We are also wired to remember novelty, to remember events that somehow deviate from our expectations. In many ways intelligence is really a measure of our capacity for prediction, whether that capacity is encoded into our brains via DNA or via our life experience. It's smart to flinch when objects suddenly loom overhead because objects suddenly looming overhead are often a sign of something dangerous about to pounce on us. It's smart to pump the brakes softly when the road feels icy because icy roads are often a sign that applying the brakes normally will send you into a dangerous skid. Intelligence is about sensing cause and predicting effect. So it makes sense that novelty should occupy such an important place in our mental apparatus. It's as though our brains contain this fundamental rule: If you're expecting X and you get Y, take note!

For as long as I can remember, I've had a strange propensity to form intense memories of certain arguments made by friends or teachers or colleagues almost in passing, over dinner-table conversation or in seminar rooms. Someone will make an offhand defense of Castro's economic policies, or the films of Jean-Luc Godard, or Madonna's latest record, and for some reason, their words will be engraved into my memory banks, and I'll find myself working through them months or even years later, building counterarguments in my head or reaffirming their essential truth with new evidence. For a long time I was puzzled about the selection criteria with regard to these memories: why was I remembering this one line so vividly, and forgetting so many thousands of others?

This pattern only started to make sense to me after I began reading about the way the brain's attentional and memory systems are designed to record novelty and surprise. All those comments trapped in my long-term memory had one thing in common: they had surprised me in some fashion. You're listening to your libertarian friend roll through his usual routine about the brilliance of Ayn Rand, and then all of a sudden he announces his support for progressive taxation. Or you think you've mastered the basics of evolutionary theory, and then someone makes a passing reference to some subfield of Darwinian thought that you've never encountered—spandrels, say, or the prisoner's dilemma. Your brain suddenly stands at attention: "Hey, what was *that*?"

You can see this mechanism captured in the wonderful French expression *l'esprit d'escalier*—literally, "the wit of the staircase"—that the *Oxford Dictionary of Quotations* defines as follows: "An untranslatable phrase, the meaning of which is that one only thinks on one's way downstairs of the smart retort one might have made in the drawing room." We haven't thought of the smart retort in the drawing room because the barb we're responding to surprised us, caught us off guard. We have plenty of good retorts handy for predictable comments; it's the ones that come out of the blue that perplex us. Sometimes we're still mulling over potential retorts on the way down the staircase because we've suffered a social slight by not being quick-witted enough to respond. But we're also mulling because our memory is designed to dwell on events that surprise us.

Researchers now believe that there is an entire neurochemical system devoted to the pursuit and recognition of new experiences and surprise, particularly experiences pertaining to reward. This system, they say, is largely regulated by the brain's production of dopamine. Because it plays a central role in addiction to several drugs, including cocaine, dopamine is often described as one of the brain's "pleasure" drugs. But this shorthand description is misleading. First, like the other major neurotransmitters, dopamine is utilized widely throughout the brain, in many areas with limited connection to pleasure or reward. (The movement dysfunction of Parkinson's disease appears to be related to reduced supply of dopamine in the motor areas of the brain.) But the problem with the image of dopamine as a pleasure drug has another level to it as well. The opioids are pure pleasure drugs—fill your brain with them, naturally or unnaturally, and you'll feel good. This is why some of life's most important behaviors—sexual climax, social bonding—trigger opioid release in the brain. Dopamine, on the other hand, is not so much a pleasure drug as a kind of pleasure accountant. It anticipates rewards that it expects the brain to receive, and sends off an alarm if the reward exceeds or falls below the anticipated level. It's not unlike what a stock analyst does in watching quarterly earnings reports: if the company meets expectations, there's no news. But if the company shows an unexpected loss or a surprise profit, then there's something to talk about. If you're expecting a certain reward—seeing the face of a loved one, landing a new client—and the reward comes through as promised, the dopamine in your system remains level. If you're denied the reward, dopamine production drops accordingly. And if the reward turns out to be even better than expected—the loved one shows up with a bouquet of flowers, the client brings in twice as much business as he had originally projected—then your brain releases extra dopamine to signal the good news. Narrative systems—movies, novels, fairy tales—exploit this drive for novelty: we like twists in our stories because our brains have a biologically grounded interest in surprise.

Lowered dopamine levels help activate what Jaak Panksepp calls the mind's "seeking" circuitry, propelling us in search of new avenues for reward in our environment. If you're expecting a three-course meal and you get a pretzel instead, your lowered dopamine levels will send you immediately to the fridge. Chronically low dopamine levels can induce the cravings of drug addiction or intense hunger—and as we saw in the last chapter, they may play a role in social addictions as well. In all these situations, though, the key recurring pattern is the dopamine system's measurement of reality versus expectation, its unwavering focus on novelty and surprise.

Addiction researchers now believe that a reason some people are particularly vulnerable to destructive habits is that their reward-expectation thresholds are easily altered by experience. Every day brings a new combination of rewards, or lack thereof. Some days are more rewarding than others. As those rewards roll in, your dopamine system assesses how closely they match the predicted levels. If your predictions are relatively steady, you're not likely to be thrown by the occasional off-the-chart day. But if your predictions are more volatile, more readily swayed by recent events, things can get more difficult to manage. Let's say one day you're expecting 5 on a scale of 1 to 10, and that day you happen to win the lottery. If you wake up the next morning expecting a repeat performance (a 10, when you're used to life doling out 5s) that day is likely to result in lowered dopamine production, because the reward won't meet your expectations. But if you possess a more stable system, and wake up the next day expecting another 5, your dopamine levels won't fall.

This is why some people try cocaine, enjoy it, and never take it again; and why some people continue taking it despite the fact that it has long since stopped giving them pleasure. Cocaine interacts with a number of different neurotransmitters, but researchers think its addictive properties revolve around its entanglement in the dopamine circuits. During the time that it remains active in your brain (usually about an hour) the drug delivers an artificial "10" to your reward-monitoring system. If you have the sort of brain that immediately resets your threshold based on the latest results, when the cocaine wears off, your dopamine levels will switch from feast to famine, and you'll find yourself craving more of the drug. (As George Carlin once said, "Cocaine makes you a new man. And the first thing that new man wants is more cocaine.") If your thresholds are less easily altered, though, the reduced reward data will just roll right off you, like the stock market ignoring a bad earnings report it had long anticipated.

The brain contains chemicals that create pleasure and reward; it also contains chemicals that create the *appetite* for pleasure and reward. Because rewards rarely just fall in our laps, the appetitive system is tied intimately to the mind's eagerness to search out new experiences. The pleasure system is anchored in the endorphins and adrenaline's near relation, norepinephrine; the appetite-for-novelty system is anchored in dopamine. These two systems often work in sync with one another, but in any individual, one system may be stronger than the other. There are hedonists, and there are seekers. The two personality types are not synonymous, though they can sometimes overlap.

Several decades ago the psychologist Robert Cloninger proposed what he called a "unified biosocial theory of personality," organized around three axes corresponding to the three major neurotransmitters: serotonin, dopamine, and norepinephrine. The serotonin axis involves harm avoidance (another version of "rejection sensitivity"). If your serotonin levels are high, you feel less vulnerable to potential slights or injuries, more confident. If they're low, you can be defensive, less willing to take risks. Dopamine, as we've seen, regulates the "novelty-seeking" axis, while norepinephrine regulates the "reward-dependence" axis, making us more or less reliant on pleasurable stimuli. Cloninger proposed that the three axes were relatively independent of one another, and that the broad tendencies of personality ultimately came down to where you landed on each of the axes. You could be heavily reward dependent, indifferent to novelty, and mildly harm avoidant—a stay-at-home hedonist, in other words. Or you could be a fearless reward-independent novelty seeker, always searching out new experiences without any real concern for whether they are dangerous or even pleasurable—a war reporter who volunteers for the frontline.

As a unified theory of personality, Cloninger's model has not yet been accepted by the psychology establishment, but as a supplement to our standard language of personality—extroverts and introverts, manics and depressives—it sheds genuinely new light, an illumination that originates in our understanding of how the brain works from the *inside*. In fact, one of the problems with Cloninger's model is that it doesn't account for the presence of other essential neurochemicals, like oxytocin or the endorphins. The ultimate problem with Cloninger's theory may not lie in its organization of personality axes around the primary brain messengers. It may be simply that he didn't include enough axes.

At some point in the not-so-distant future, we may have tools—either diagnostic tests or brain-imaging studies—that enable us to create accurate neurochemical portraits of ourselves along multiple axes. We'll be able to say with real confidence that we have unusually high serotonin levels, a dopamine system that is easily reset, slightly less testosterone than the average male. This portrait will look something like the old hit-point system devised by the creators of Dungeons and Dragons, whereby your character would have 15 points for dexterity, 12 for charisma, and 7 for wisdom.

Neurochemical profiling sounds like something out of the film *Sleeper* or *Brazil*, but it's not quite as crazy, or as sinister, as it sounds. For one, it would not wed you inexorably to the fate of your genes, since life experience and learning also alter your neurochemistry. You can be high serotonin because you were born that way or because of your upbringing. Profiling would certainly be a crude simplification of your personality, but probably less so than SAT scores or the text of a personal ad. You'll still always be able to learn more about people by hanging out with them for long periods of time, but when you don't have long periods of time, knowing something about their brain chemistry might be informative. When people flinch at the idea of neuroprofiles, it's usually because they imagine this analysis replacing all the other ways we understand personality. But it's not an either/or proposition.

We may well get to a point when we can identify our good friends based on a shorthand description of their average neurotransmitter levels. ("High serotonin, low dopamine, medium estrogen? That sounds like Carla!") Would this describe the person fully, capture his or her essence? Of course not. But it might well be more revealing than describing someone as a six-foot-three male, 142 pounds, and a firstborn child. You might easily identify a friend based on that description, and the information wouldn't be irrelevant to understanding him. It just wouldn't be the whole story. The same goes for your neurotransmitter profile. It's relevant data, once you understand what it means. It's not the whole story, but it's surely *part* of the story, and excluding it arbitrarily from our personal narratives simply because it isn't comprehensive makes as little sense as omitting our childhood experiences because they can't explain 100 percent of our grown-up selves.

The added benefit of talking about neurotransmitter profiles—as opposed to genetic profiles—is that the neurotransmitter route leaves room for the impact of life experiences and culture. Your brain chemistry is partly shaped by your DNA, of course, but it is also very much an imprint of your upbringing. A famous study years ago looked at stress levels in that most hierarchical of cultural institutions—the British civil service—and found that one's rank in the hierarchy was a clear predictor of heart disease, a telltale sign of elevated cortisol. The higher you were in the pecking order, the study implied, the lower your cortisol levels. Even the most dogged biological determinist wouldn't argue that people with innately lower cortisol levels were naturally rising to the top of the heap. Clearly the cultural environment of the civil service was affecting the stress levels of its members, and changing their brain chemistry accordingly. More status, less cortisol; less status, more cortisol. Such hormone levels would appear on one of our imagined neuro-transmitter profiles, but they wouldn't necessarily suggest some fate sealed in the double helix before birth. In fact, they might well point to an imbalance outside the individual body, in society itself. The drugs flowing through our bodies and our brains can tell us a great deal about ourselves, but not just the biological selves we were born with. They are also symptoms of a wider world outside the brain, a world that the brain's inner chemistry reflects.

I suspect that not too long from now we'll see charts of average cortisol levels—alongside those of the other major endogenous drugs—across national populations, tracked over long periods of time. Those charts will be like a macro version of the first images I saw of my adrenaline levels, each spike representing a joke I'd cracked. You'll see surges of cortisol after terrorist attacks and recessions; serotonin spikes during bull markets. World events won't alter DNA during those periods, at least not at speeds that register on the ten- or twenty-year scale. But those events will have a marked effect on brain chemistry, operating on the scale of both of seconds and decades. The brain's endogenous chemicals have always played a silent, but crucial, role in the long drama of human history. Now they have a voice.

6

Scan Thyself

In the summer of 1933, while swimming at the French vacation town of Saint Jean de Luz, the composer Maurice Ravel suffered a stroke. Although he had remained productive through a decade of mental agitation, battling depression, insomnia, and temporary amnesia, the stroke marked a turning point for Ravel. He first noticed its effects that day on his motor control, as he struggled to stay afloat in the water. But as the days passed, a more troubling long-term deficit became apparent. The stroke had destroyed his ability to create music.

Silencing one of the world's great composers was cruel enough, but the stroke had a more fiendish twist: Ravel could still appreciate music as vividly as before, and indeed his mind was filled with new musical ideas. But he had lost the ability to translate those ideas into a language that the external world could understand: either by writing or performing. In a sense, Ravel's stroke had left him with the reverse of Beethoven's legendary deafness: he could take in music from the external world, but he couldn't give it back. "I've still got so much to say, so many ideas in my head," he would lament to his acquaintances. But those ideas were trapped inside the brain's black box, where they remained until his death in 1937.

Because Ravel's stroke also greatly impaired his mastery of written language—his biographer describes him taking eight days to compose a fifty-word letter to a friend—neurologists now believe that the composer had experienced a left-hemisphere stroke, damaging the linguistic centers in that part of the brain while leaving the more emotional right hemisphere intact. Music could still move Ravel after the stroke, but he couldn't translate that passion into symbols or physical movements; he could hear the totality of the musical sound, but couldn't break it down into its component parts.

Ravel's stroke reveals a typical pattern in the way the brain processes musical information: ordinary listeners generally rely on the right hemisphere when enjoying music, while musicians—particularly those capable of reading and writing sheet music—show additional activity in the left hemisphere. Ravel's musical aphasia supplies yet another example of the mind's fundamental modularity—even a seemingly unified task like composing music turns out to involve specialized areas of the brain: one hemisphere for dreaming up the melody and harmony, and one for transcribing them.

When we talk about musical genius, particularly among composers, what we're normally celebrating is a fusion of left- and right-hemisphere accomplishment: taking the intangible passions of music and turning them into something that can be recorded, transcribed, passed on to other musicians and other ears. Most of us ordinary listeners have to make do with the simpler, right-hemispheric pleasures of enjoying other people's music.

But if you think about our pleasure in music from a distance—

think about it the way we've thought about tickling or mindreading—it becomes a strange convention. Enjoying music seems simple compared to the notational skills of the great composers, but
that simplicity is deceptive. Why do the raw wave forms of music
have such control over our emotions? We feel passionate about our
kids because that passion helps them survive in order to pass on
their genes. But why do we feel passionate about a ballad or a guitar
riff?

The more I understood about the brain, the clearer it seemed
that the science could teach us immense amounts about behavior
that triggers dedicated circuits in our heads: paying attention,
falling in love, being afraid. These are all regions of experience that
have unmistakable evolutionary significance, so it's no surprise that
we should find specific architecture in the brain corresponding to
each of them. But life is more than just instincts, and some of
humanity's great pleasures come from experiences that seem, on the
surface at least, to have a less direct connection to our evolutionary
past. I know why I feel such a powerful bond with my children, but
I have a hard time explaining why I still feel chills down my spine
when I listen to Van Morrison's *Astral Weeks*, even though I must
have played it a thousand times. What light could brain science
shed on that mystery? Science has much to teach us about our
instincts, but what about the intangibles?

One potential route involves changing the terms of the question. Instead of asking *why* music moves us, we can ask something
else: what happens in our brains when music moves us? We may
never know the evolutionary explanation for music's hold on the
human psyche, and indeed there may be no direct explanation
available, in that an ear for music may not be a trait that was directly
selected for. (Music may be one of Stephen Jay Gould's famous
"spandrels"—indirect by-products of other selected traits.) But we
already know something about what actually happens in the brain

when we enjoy music. As Ravel's stroke demonstrates, most appreciation of music happens in the right hemisphere, which suggests that the intuitive opposition between language and music, between concrete categories and the more fluid associations of sound, have their origins in the brain's bilateral architecture.

We also know something about that most elusive and private of music experiences: the chill. Jaak Panksepp has been in pursuit of the neurochemistry of musical chills for more than a decade. His work—now supported by a number of other studies—makes a convincing case that the shiver of pleasure we experience while listening to our favorite music is the release of endogenous opioids, the same molecules implicated in social bonding, parental love, the "runner's high"—and, of course, in narcotic drugs like heroin and morphine. Panksepp has found that animals appear to have chill responses to music as well. In one widely cited study, he played dozens of records to chickens attached to equipment designed to record their shivers of pleasure. (The chickens turned out to have the strongest positive response to the late-era Pink Floyd record *The Final Cut.*) Here, again, a little knowledge of brain chemistry illuminates a new twist on our most familiar experiences: the pleasure of listening to music strangely connected to the pleasure of parenting, or of taking illegal drugs.

Imagine, then, taking Panksepp's experiment one step further: instead of a chicken's brain listening to Pink Floyd, let's peer into Ravel's brain—or one like it—as it dreams up a new composition in the years before his stroke. Thus far, most brain-imaging research has focused on normal brains and on brains that suffer from some kind of disability. But we also have the opportunity to scan brains that are unusual in the sense of being unusually gifted. What vista into the world of inspiration will this open up to us?

I don't know firsthand what moments of true musical inspiration feel like. For me, inspiration revolves around words and sentences, and not melody and harmony. I'm not imaging myself to be a literary Ravel, but stringing text into narratives and arguments has been the most fluid of my mental faculties for as long as I can remember. Could brain science have something useful to say about this talent? I wanted to know what was happening in my head when a new insight arrived, usually half formed and barely grasped: a vague connection between two ideas, a new way of introducing a troublesome chapter, a phrasing for a sentence. This faculty was less charged emotionally than many of the experiences I had explored in writing this book, but it was no less mysterious to me. For reasons probably both genetic and cultural, I am not much of a mystic, but these flashes of insight were the closest thing I had to the experience of mysticism. These sparks were the transcendence that Keats sought when he commanded us to "open wide the mind's cage'd doors." An idea shoots in front of my mind's eye seemingly out of nowhere. Where did it come from?

How extraordinary that we can even begin to answer this question! We can only speculate where new ideas come from in the sense of their evolutionary roots, and we don't really understand how the firing of neurons creates the rich subtleties of ideation. But we can determine, with split-second precision, the parts of the brain that are active in the creation of a new idea. We can map mental processes as ephemeral as having a hunch. On a fundamental level, we can tell *where* the hunch comes from. All it takes is a brave, nonclaustrophobic subject and a $2 million magnet.

I thought I was precisely that brave, nonclaustrophobic subject until they strapped my head down to the mechanical gurney, and I began sliding into a two-foot-wide tube, with only a mirror the size of a playing card supplying me with a glimpse of the outside world.

There's no better way to say it: I was having my head examined.

Mechanically speaking, the exam was being conducted by a five-ton GE Twin-Speed fMRI scanner. My guide through the world of advanced brain scanning was Joy Hirsch, director of Columbia University's Brain Imaging Group, who had graciously offered to help me in my pursuit: to see the brain, from the inside, as it comes up with a new idea.

In a sense, this pursuit had begun after my original experience with the Attention Trainer's neurofeedback device. A few sessions analyzing my beta levels during various attention-related tasks had made me curious about my brain's behavior during other activities. Writing itself has a strange hold over my psyche. I can be totally exhausted at the end of a long day, without the slightest urge to work, but if you sit me down in front of a computer and pull up a piece that I'm in the middle of writing, I'll invariably start tweaking the text—rearranging a phrase, inserting a few qualifications, punching up an opening line. It feels almost like a compulsion; I can't *not* tinker with the words. Before our son was born, I worried that it would be difficult to write with a toddler charging around the house, and indeed I found that many other tasks that required concentration—reading, interviewing someone on the phone—were in fact quite difficult with our son in the room with me. But writing was a breeze. When I'm truly locked in working on a passage, a 747 could be taking off in the room and I wouldn't notice.

So this was what I had come to Joy Hirsch to understand: when I'm in that zone, what is happening in my brain? I knew Joy was probably as well equipped as anyone on the planet to capture that mental activity; her center had just installed the state-of-the-art fMRI machine, and she had decades of experience interpreting brain images. The question was whether we could construct an experiment that would reveal this activity as clearly as possible. Would it even be possible for me to come up with an interesting idea with my head stuck inside a five-ton magnet?

A week or so before my appointment with the scanner, I suggested an experimental structure to Joy: we would begin with my reading a series of nonsense sentences, followed by my reading someone else's prose, and then I would read a passage of my own work—a passage from this book, in fact. In reading my own passage, I hoped to spur one of those imaginative leaps: something about the words would make me think of a new line to add, or a new way of phrasing the idea, or some other unpredictable insight. If all went well, the machine would take a snapshot of that idea forming in my head. Unlike neurofeedback technology, fMRI scans can capture subtle shifts of activity within a three-dimensional model of the brain by measuring levels of oxygenation in the blood of nerve cells. It is not a perfect view by any means—you have to have roughly 500,000 neurons active in an area for the scan to register them—but it is as close to pure vision of the mind's inner life as current technology allows us.

When I arrive for my session, Joy and I sit down in her office to review the terms of our experiment. She begins by telling me that she's replaced the initial "control" experiment of reading nonsense sentences with a standard visual test pattern of a flashing checkerboard.

"You can't really use nonsense as a control, because the brain goes nuclear with nonsense," she explains. "With normal activity—reading, or touching an object, or recognizing a face—you see very predictable activity in specific regions. It's like the brain is handing off the task to the appropriate area. But when you have noise, the whole brain seems to light up trying to make sense of it." There was something lovely in that image: the brain, faced with apparent chaos, leaning on all its resources looking for some hope of order in the mix.

Joy explains that each stage of the experiment will involve three sections of forty seconds each: rest, activity, rest. The scanner will start up, and I'll do my best to think of nothing for forty seconds. Then the stimuli will begin—the flashing checkerboard or the text—and I'll process that for another forty seconds. And then I'll think of nothing again. Each 120-second stage will be repeated twice.

As Joy lays out the sequence, I start to worry that I won't have time to actually *think* while in the machine; I don't want to spend the whole forty seconds reading, particularly once we get to my own words. I want to have the words trigger some new idea or association in my head. So Joy agrees to make a last-minute addition: a final stage during which I'm shown a single sentence from my book and given the entire forty seconds to ruminate.

Then Joy walks me through the risks. "We're looking at your brain here. So there's a very small chance that we might see something in these scans, some abnormality."

I nod. "You mean a brain tumor."

"Sometimes when we do work with experimental subjects— people who come in to help with our research, and who don't have any symptoms—they say, 'If you see something in there, don't tell me.'"

"Hey, if you see something in there that you don't like," I smile ruefully, "by all means let me know."

Then she moves on to the dangers associated with the scanner itself. "It is a fundamentally safe procedure, noninvasive." I think of a news story from a few years back in which hospital staff had left a metal trash can in the room with an fMRI. When they began scanning a patient, the magnetic field triggered by the scanner being switched on turned the trash can into a lethal projectile that killed the guy instantaneously.

I choose not to bring this up.

Then her voice turns slightly more serious, which makes me think that whatever she's about to relate is something she deals with more regularly than tumors or flying trash cans. "You should also know that some people find being inside the scanner uncomfortable."

"Because it's so claustrophobic?"

"It's a small space, and the machine makes a lot of noise. Some people have a hard time in there. But you should know that I'll be there in the room with you, and if at any time you want to come out and catch your breath, we can do that very easily."

"I think I'll be all right," I say more or less honestly. I have my fair share of fears, but confined spaces isn't high on the list, and as long as no flying metal objects hurtle toward me in the first few seconds, I suspect I'll feel pretty safe in there.

A minute or two later, we walk over to the fMRI room. The machine itself looks like an oversized clothes drier—about ten feet high with a huge GE logo embossed above the hollow tube at its center. I lie down on the mechanical gurney, and the technician gently tapes my forehead to the cradle at the end, hands me a pair of earplugs.

And then I'm in.

Being inside an fMRI machine is definitely more unpleasant than it looks to be from the outside. The space itself is astonishingly small, and the sense of being encased in a huge piece of machinery unsettles more than you think it will. For my experiment Joy and her team have placed a small mirror above my eyes that enables me to see a sliver of the world outside the tube. This sliver lets me read the text that they've projected onto a screen, but it also prompts a surge of nausea as I first enter the scanner. That queasiness, I know, is yet another side effect of two modules sending conflicting information: part of my brain reports that I've just been inserted in a cramped tunnel, while my eyes report a clear vista across the room. For a sec-

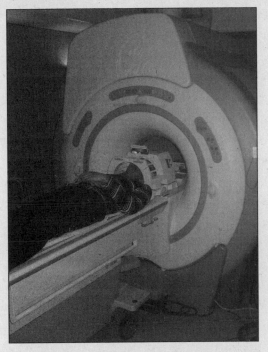

The author during his fMRI exam.

ond, I think, *I may actually have a problem with this. I may be one of those people who call out for a pit stop.*

And then I do what I normally do in a stressful situation, what I did on the biofeedback practitioner's couch. I make a joke—an internal joke that only I get to appreciate, but a joke nonetheless. I think to myself: *How is it that I ended up here? What strange series of life events led me to the point where I'm actually* asking *to be put into this insane device?* And after that, I'm all right. Uncomfortable, but all right.

The fMRI machine is capable of capturing two types of images: conventional MRIs that are higher resolution but don't show specific activity in the brain, and then lower-resolution "functional" images that show the brain actually thinking. (Functional MRI images work because active areas of the brain require an increase in oxygenated blood, which creates a small but detectable disturbance in a magnetic field.) We begin with a round of conventional images of my brain, during which time the machine rattles ominously around my head. Then we move on to our little experiment, starting with the checkerboard pattern.

You can easily tell when the fMRI is in its "functional" mode because it emits an uncomfortably loud, high-pitched, pulsing tone. (Hence the earplugs.) When you're actually inside the scanner, it sounds like a truck backing up into your head. For the first forty seconds of "rest," I find myself incapable of thinking about anything other than the excruciating noise. When the flashing checkerboard appears on the screen, it occurs to me that this is like attending some kind of demonic performance-art happening—a tiny, cramped space with strobing black-and-white images projected onto a screen, all accompanied by monotonous, piercing rhythmic tones.

But by the second iteration of the checkerboard stage, I start getting accustomed to the noise and the physical enclosure. I can see Joy smiling at me through the mirror, and the sound becomes more background noise than anything else. In fact, I feel comfortable enough that I start having difficulty shutting off my brain during the "rest" periods. First, I find myself thinking about ways that I could describe the setting, shaping the story of my fMRI experience while my head is still stuck inside the device. When I catch myself doing this, I smile in my dark tunnel. It occurs to me that this is one of those small examples of the brain's miraculous resilience and flexibility: you stuff your brain into a physical situation that should by all rights overwhelm it, *and* you tell it explicitly not to think of anything, and yet still it churns away in spite of everything. You couldn't imagine a more hostile environment for free associating, but here my brain was riffing away, as though I were daydreaming in the shade of an oak tree.

Then I'm reading. Processing text turns out to be a bit of a strain, given the whole rearview mirror apparatus. Joy had selected a couple of passages from Nobel Laureate Eric Kandel's classic neuroscience textbook, while I had sent in a few paragraphs on Freud from an early draft of this book. I have to force myself to actually read the Kandel text, and not think about the bizarre setting. Of course, as anyone who ever suffered through fifth-grade mandatory summer reading assignments knows, when you have to actively remind yourself to pay attention to what you're reading, you're usually not reading very closely. Scanning the Kandel as it's projected onto the screen, I have to fight to keep up with the text. (If I'd been tested afterward, I'd wager that my retention would have been less than 50 percent.) It ends up being easier to focus on my own words, but there certainly isn't time to ruminate. As we finish that stage, I think to myself that I'm glad we added the rumination "bonus round."

I'm glad, but I'm also getting tired. I haven't moved my head more than a centimeter in around twenty-five minutes, and the space is starting to close in on me. When the first frozen slide of text arrives on the screen for the rumination stage, I feel like I've been caught off guard. "Shit!" I say to myself. "Now I have to think of something." For forty seconds of this $2 million machine's time, I think of absolutely nothing worthwhile. I think about trying to think about something. If there is a cognitive version of flailing, this is what I do for the first scan.

But when the second round—the last run of the entire experiment—arrives, I'm prepared. I decide to let my brain do what had come naturally to it throughout the experiment. I've already started down the road of describing the experience in the scanner—why not take this last round and actually start working out the language? And so when the text flashes up on the screen, notifying me that the forty-second rumination period has begun, a sentence starts to take form in my head. I am writing.

The words I string together in the fMRI are roughly the same words you encountered a few paragraphs ago describing the resilience of the brain in the most uncomfortable of situations. The general idea arrived a few minutes earlier, but the exact phrasing originates in that last session. The specific sentence, of course, is incidental; what makes it interesting is that Joy Hirsch and her fMRI are watching as it forms in my head, as my brain pulls the words out of the nothingness and makes them into something fixed—sturdy enough to remain intact until I sit down at my computer several days later to type them. For those last forty seconds, I have stumbled into my own small version of the zone—the one I have been wondering about since my first round with the Attention Trainers. And the cameras are rolling.

The results arrive in two stages. The first stage comes almost immediately: Joy gives a quick glance at the conventional MRI images of my brain, and announces that I have a healthy specimen. "Everything looks great," she says as she slaps the X-ray-like film onto a light board. "A textbook brain." I glow with pride for a second, and then think, *She probably tells this to all her experimental subjects.* Still, I find myself more pleased than I had expected to find out that I have no visible brain tumors. I think, *At least I've got that going for me.*

The second stage is where it gets interesting. A few days pass, and Joy sends an email to let me know that the results are in. "You're going to like this," she writes temptingly. The next afternoon I take the A train up to 168th Street, and Joy and I sit down at a conference table to spend some quality time with my brain.

Joy has assembled a collection of about forty color printouts, each displaying four images of my brain at work. The images are overhead views, and each one is a "slice" of my brain, starting with the brain stem, at the very bottom, and ending with the tip of the cortex. For each stage of the experiment—there are four in total—the fMRI has captured twenty-five slices of my brain going about its business. That business takes the form of changes in blood flow to different regions; the scanner first looks at my brain during the "rest" periods, then during the "activity" period, and it records any salient differences between the two. These images let you see the areas that are relevant to a particular task, and shut out the background processing that the brain is always doing. My brain stem, for instance, was steadily plugging away maintaining my breathing pattern—along with many other mission-critical operations—but that area doesn't light up on the scan images because those patterns didn't change during the experiment.

Areas that do show noticeable changes appear on the images as a cluster of bright yellow pixels, fading out to orange and red at their peripheries. The images look strikingly like the Doppler radar images you see on the Weather Channel. (If you blur your eyes a little, you might think that yellow patch on the image was a thunderhead, not a brainstorm.) The image is projected over a grid with numbers running along each axis. The numbered grid and the slices create a three-dimensional system of coordinates, the latitude and longitude of neuromapping. The grid is made up of small cubes called "voxels," and each voxel has a specific address. (My amygdala is located at voxel 65, 70 of slice 13.) This lets you make easy comparisons between activity in scans of different brains, as well as look up areas in an artful, hand-drawn brain atlas that Joy consults at various times during our conversation.

Joy begins by laying down the twenty-five slices for stage one of our experiment, the dreaded checkerboard. The pattern of activity is immediately visible, even to my untutored eyes, mostly because there's literally nothing going on in 95 percent of my brain. Only a thin band wrapping around the back of my head, roughly at ear level, glows yellow.

"We know that the flashing checkerboard is a very salient stimulus for just the visual processing areas of the brain," she says. "And that's exactly what's happening here."

She points to the yellow band: "This part of the brain is all primary visual cortex. What's unique about this is that this activity doesn't get out of the occipital lobe—and *nothing* goes on in the frontal lobes. Nothing. This is just as exclusively visual as you can get." We both start to laugh. "Your brain is doing the minimal amount it has to do to sit there and look at that stupid checkerboard!"

Looking at those blank areas on my mental map reminds me of all the times that someone had gravely explained to me that we only use 10 percent of our brains, and then waxed rhapsodic about how smart we'd be if we could tap 100 percent. Of course we only use a small percentage of our brain at any given time—and it's a good thing, too! Your brain has dozens of dedicated tools, most of which aren't relevant to whatever it is you're focusing on right now. If your visual cortex keeps kicking into overdrive as you're trying to memorize a speech, the words won't stay in your head as readily. Only using 10 percent of your brain is a sign of *efficiency*, not underachievement. Arguing that we'd be better off with 100 percent is like raving about how great Shakespeare would have been if he'd managed to use all twenty-six letters in each of his words, instead of a small fraction of the alphabet.

Joy lays down the slices from stage two, when I read the mystery passages from Eric Kandel. The contrast with the first-stage images is startling: while the back of my head is lit up similarly, there's much more activity in the rest of the brain. Joy says, "You expect to see some visual processing, of course, because you're reading. But we'd also expected to see some higher-level functions."

She starts by pointing to a pair of yellow clusters, aligned symmetrically on the left and right sides of slice 12, about halfway up. "That's an area associated with eye movements—your eyes are darting back and forth reading, which they weren't doing in the checkerboard stage."

"Now, look at the difference here," she says, pointing at the slices one level up. There's quite a bit of activity, both in the middle of my brain and at the peripheries. "This is when we can see the higher-level processing. This is definitely language related—the dorsal part of Broca's area. There's just a lot more frontal activity all around." She puts down another set of slices. "This is Wernicke's area, loud and clear, whereas in the checkerboard phase, there's nothing there, just nothing. So your language areas, the visual system, the eye movement area—they're all participating in your reading of the text."

I need a little help in locating the major landmarks, but once I've got my bearings, the pattern is unmistakable. I feel a little like an autistic person learning to read facial expressions: it doesn't come naturally, but you can do it with enough practice. I turn to Joy and say, "So if you knew nothing about this experiment, and just looked at these images, would you be able to tell that this was someone reading?"

"Absolutely, absolutely. It's a textbook case." Then she smiles mischievously and starts to lay down the slices from stage three. "And this one—I'd say that this one was someone reading his favorite author."

We're looking at my brain reading my own words. At first glance, the images roughly parallel the pattern of the previous stage, but they're much hotter, as though the current has been turned up. The yellow clusters are larger and more pronounced. "Eric Kandel—Nobel laureate or not—can't hold a candle to this," Joy says, breaking into laughter.

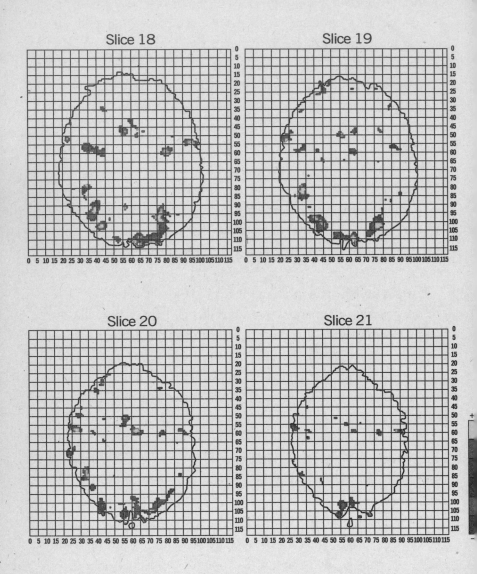

The Checkerboard Test

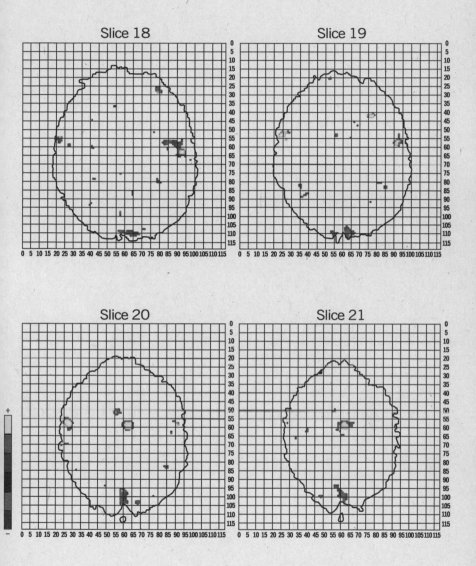

Reading

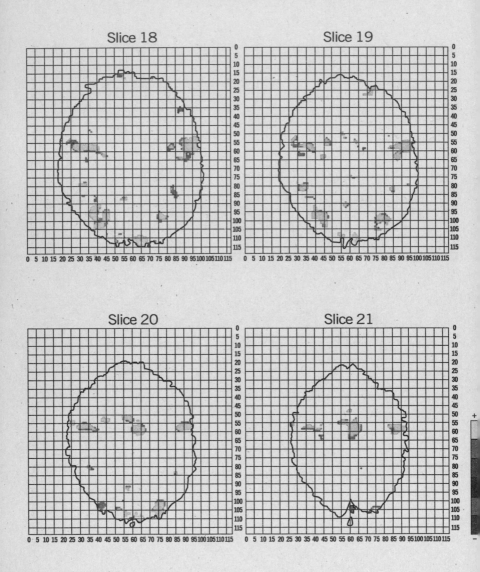

Failed Rumination

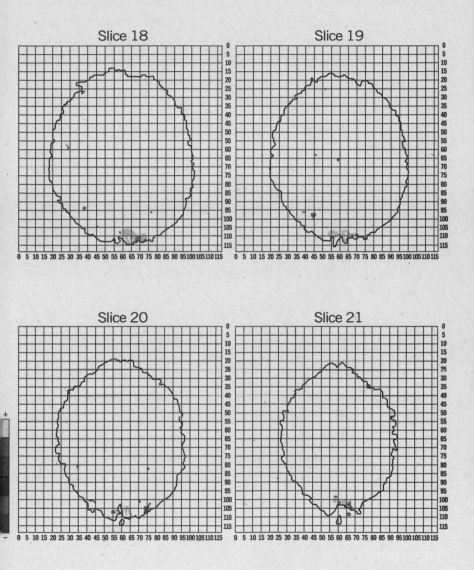

Slice 18   Slice 19
Slice 20   Slice 21

Successful Rumination

"Oh dear." I chuckle, slightly embarrassed at having dreamed up the experiment. "It was a vanity project from the beginning."

"Look at this," she says. "The same areas are working, but they're working much harder with your own words. It's amazing."

"I'm all aglow." I shake my head as I take in the images. I note that the hippocampus—the seat of memory—is now burning brightly, where it had been a dull red in the Kandel round. "So I've got more associations being triggered by these words because I wrote them."

"That's exactly right," Joy says. I think of all the times I've complained that it's hard to get a good feel for your own prose in its published form, because you've been there for all the first drafts and false starts, all the edits and tweaks and substitutions. All those alternative sentences crowd out your present-tense experience of reading. Now I can see that crowding directly, traced in those yellow voxels on the page.

I suppose it's possible to see this moment as the ultimate exercise in postmodern hall-of-mirrors self-reflection: you, dear reader, are reading a book describing a brain reading the book you're reading being read by a $2 million magnet. Who needs *The Matrix* when this is reality? Yet I think it's more accurate to see the activity as exactly the opposite: not an endless series of reflected reflections, but instead "dartlike and definite"—seeing the brain's actions directly, prying the mind open and taking a good look. I can see my hippocampus lighting up, filling my brain with associations and trace memories as I read my own words. That's reality, not illusion.

There's something in Joy Hirsch spreading out the images on the table that brings to mind a tarot card reader, but there's nothing mystical in her analysis. I find myself thinking, *This person I barely know has ventured inside my head in a way that no one has ever ventured before.* That's why the hall-of-mirrors interpretation feels wrong to me. It's not an endless simulation I've entered into here, but rather something that feels authentic, even intimate.

Thus far all the images we've examined have been composite sketches: each stage included two runs, and so the images are a combined look at activity over the two of them. But with the rumination round, I had asked Joy to look at the two runs separately, because I had fared so poorly the first time around and because in the final run of the day, I had managed to get my brain exactly where I'd wanted it to be for my forty seconds in the spotlight.

The images from those two sessions do not disappoint. In the first run, small spots of activity are scattered across my brain, mostly in red voxels (suggesting less activity than the yellow). There's little shape or symmetry to the map; my brain looks cluttered. But in the second run, what jumps out at me immediately is how silent most of my brain appears. Only the language centers light up with any intensity, along with a sharp yellow rod at the center of my brain, extending up to the very top of my cranium. There's very little visual activity, and almost nothing from the eye-movement regions.

"There's a concept of efficiency that has emerged in the neuro-imaging community in the last few years," Joy says. "It's basically that when there's a task that the brain is having difficulty doing, the pattern looks very distributed, like this here." She points to the cluttered image of run number 1. "This was not an efficient action—as opposed to here, where the specific tools of the brain are contributing in an efficient way to the task at hand."

"You really look like you got your act together here." She's point-ing to that bright yellow dot on the upper images of run number 2. "Here's more evidence of that—look at this very focused medial frontal gyrus. This is one of the most distinguishing characteristics of this scan—this is a very high-level executive function of the brain, and you can see it running like a pole all the way down to the cingulate. I think that the medial frontal gyrus is important in coordinating different activities in the brain, reaching for the right tool at the right time. In this last scan, the entire structure—not just a part of it—is active." In Joy's phrasing, my language areas were perfectly "robust" during these inspired forty seconds, but they didn't turn out to be the most interesting element of the image. It was the overall orchestration, the clarity of the pattern, that stood out, the lack of mental clutter.

What had I been hoping to find? I thought about this on the subway ride home. In the crudest sense, I suppose I thought that my skill at stitching words together in my head might turn out to have its own modulelike presence in the scan: a distinct patch of neurons devoted to imagining sentences. If the brain is filled with all these modular tools, then somehow it seems logical that tasks you're good at should have some visible presence on the brain map. Sometimes this is the case: Einstein's brain had unusually large inferior parietal lobes, which we think gave him his extraordinary spatio-logical skills. (He famously solved problems as images in his head weeks before he could turn them into working equations.) Such a skill most likely would have shown up directly on an fMRI: a person gifted in spatial intelligence shows more activity in regions of the brain dedicated to spatial processing. I suspect that left frontal lobe of Ravel's brain would have lit up brilliantly had he been able to take an fMRI before his stroke.

But in my case, the scan revealed something quite different. (I'm no Einstein, as it turns out.) There was no special module. What caught Joy's eye in the final rumination scan was not a specific region, but the overall *pattern* of brain activity. The tools in the toolbox weren't particularly impressive, but the toolbox itself was well organized. In fact, the only specific region that seemed to be at all above average was the one responsible for coordinating activity in other regions. My language areas were perfectly adequate, and my hippocampus seemed to kick in nicely when I was engaged with interesting text (or at least my own text). But perhaps the most telling thing about my brain map was what didn't show up on the images: when I was focused, there was almost no activity in areas that weren't related directly to the task at hand. Compare that to my episode of cognitive flailing in the first run of the rumination stage: on that scan, there's hardly a discernible pattern. It's mostly noise, and little signal.

I have no idea how replicable my fMRI results would be if I tried the exact experiment again, and it's unclear whether that pattern of organization—with its strong medial frontal gyrus and its many silent regions—holds true for my brain generally, or just for this little snapshot. But I suspect there is a larger truth nestled in that last fMRI image, one that has begun to change the way I think about people I know, much as learning about mindreading transformed the way I thought about people's social skills. I suspect that the world of talent is made up of two kinds of brains: some that have specific modules that are unusually good at their job, and some that are unusually good at keeping all the different modules organized. Both types of brains come across to us as talented, as intelligent, but I think the types are different enough that you can learn to recognize them if you know what to look for. We all know people who have dazzling skills: they can sit down at a piano and pick out a tune they heard last week; they can calculate interest rate payments in their head; they can actually understand quantum mechanics. But we also know people whose brains seem gifted in a different way: no stunning, off-the-chart skills, but a general competence and efficiency, with very little noise complicating their signal.

My dad used to say to me during my high school years: "You're not a rocket scientist, but you're smart and you've got a lot of talent." I used to bristle at the remark. (If I wanted to, maybe I *could* be a rocket scientist!) But now I think he was onto something. I've met rocket scientists—and astrophysicists, and programming wizards, and architectural geniuses—and I don't possess anything like what they've got mentally. I don't have their special gifts. But those fMRI images made me think that perhaps I have something else, a little less dazzling, but nothing to be ashamed of either. Maybe I have a well-*orchestrated* brain—with no world-famous soloists but a nice sound nonetheless. In a sense, this is what my dad had been trying to say, in slightly different language: I was talented in an orderly brain kind of way, not a supermodule kind of way.

It was only one experiment, but the machine had given me something that machines don't normally deal out: a hunch about myself, and maybe a larger hunch about people in general. I'd been dreaming for more than a year of capturing my brain as it came up with an idea, and thanks to Joy and her uncanny device, I'd managed to catch precisely that glimpse. The results were mesmerizing and remarkably legible, even to my untrained eyes. But they didn't provide unequivocal answers or magic bullets. They were more like clues.

Those fMRI scans of my brain were, technically speaking, the end of my journey inward—but they felt like a beginning. Seeing my brain come up with an idea had given me another, more interesting idea, one that still reverberates in my head as I write. Wouldn't it be nice to have a scan of *that*?

CONCLUSION
Mind Wide Open

"The deficiencies in our description would probably vanish if we were already in a position to replace the psychological terms with physiological or chemical ones. . . . We may expect [physiology and chemistry] to give the most surprising information and we cannot guess what answers it will return in a few dozen years of questions we have put to it. They may be of a kind that will blow away the whole of our artificial structure of hypothesis."

—FREUD

All of us walk around with an operative theory of how the mind works. It's rarely a unified theory, of course: typically our models are cobbled together out of different disciplines and intellectual periods. We'll dabble in Eriksonian psychology and say that someone is having an "identity crisis;" we'll borrow from modern neuroscience and describe ourselves as "very right-brain"; we'll steal a page from the mystics and refer to the Jungian unconscious or the personality traits revealed by astrology. But while our popular theories of the mind are mostly mongrels, they invariably share one common ancestor: Sigmund Freud.

Freudian assumptions about how the mind works remain ubiquitous in our culture—so ubiquitous, in fact, that we seldom even think of their original provenance. Freud's ideas are like coins that have been so long in circulation that the insignia stamped onto their surface has worn off. When you allude to repressing a distasteful memory, or joke about a revealing slip of the tongue, or you talk through your memory of a traumatic event to lessen its hold over you, or analyze a friend's dream for its hidden meaning—when you do any of these things, you're speaking the language of Freud, using a grammar of psychological categories and relationships that he largely invented.

The argument of this book has been that modern neuroscience presents us with a new grammar for understanding our minds. You don't need a Ph.D. to speak this language; with the right tools, and the right translations—some of which I've attempted over the preceding pages—you can get to a level of fluency that will make you a more informed, more self-aware inhabitant of your own head. For a hundred years, much of Western society has assumed that the most powerful route to self-knowledge took the form of lying on a couch, talking about our childhoods. The possibility entertained in this book is that you can follow another path, with equally insightful results: going under the fMRI scanner, or hooking up to a neurofeedback machine, or just reading a book about brain science.

If you spend some time exploring this new world, you will end up with a set of conceptual building blocks to use when thinking about how your brain works: some of them specific chemicals, some of them localized regions, some of them broader patterns of interaction between regions or chemicals. A handful of these categories have been trickling out over the past few decades: the left-brain/right-brain meme, the natural high caused by endogenous opioids, serotonin's ties to social confidence. Expect a flood of them over the next decade. Thanks to the anxiety-prone nature of life since 9/11, the amygdala now teeters on the verge of becoming a household term. Google now reports that 103,000 pages on the Web mention oxytocin. As I was writing this book, I attended a meeting with a leading American politician, the topic of which had nothing whatsoever to do with neuroscience. When I told him that I was writing about the brain, he looked at me knowingly and said, "It's all in the limbic system."

But if this new language is poised to transform our popular assumptions about how our brains work—if the *politicians* are starting to speak this language—the key question becomes: what will happen to the old language? Are those Freudian categories obsolete in the age of neuroscience? Or is it simply a matter of translating the old categories into a new tongue, trimming here and there where necessary? Given everything we know about the brain's inner life, what parts of Freud are worth keeping? And more than that, what parts can we still learn from?

For almost a hundred years, we've been locked in a relationship with Freud that is probably best described, using language that Freud himself helped create, as bipolar. Only Marx's track record exceeds that of the Viennese doctor in terms of mass embrace followed by renunciation. For fifty years, Freud's model of the psyche towered over models generated by his competition; his dense, literary analyses spawned an amazing number of household terms: Freudian slips, Oedipus complexes, wish fulfillments, dream cigars that are, alas, just cigars. And then just as quickly, Freud was under attack from what seemed like all quarters: from the pharmacologists, who found in lithium a far more effective treatment for manic depression than any talking cure; from the behaviorists, who tried to turn psychology into a science of external action, away from internal mental life; from the feminists, for whom Freud's case studies on child abuse seemed like exercises in blaming the victim; from the neoconservatives, who considered the good doctor uncomfortably sex-obsessed and who found the whole idea of divided selves and unconscious drives laughable; from the brain scientists, who had begun to peer into the brain's inner geography using imaging tools and had failed to detect the kingdoms of id, ego, and superego.

After decades of mood swings, we may finally be winding our way back to an equilibrium point, where we can recognize the extraordinary conceptual breakthroughs that Freud made possible while still acknowledging that some elements of his theory need updating in the light of modern brain science. Perhaps the most interesting sign of this emerging balance has been the rise of the neuropsychoanalysis movement, spearheaded by an eclectic group of brain scientists and psychoanalysts intent on exploring the ways in which the modern understanding of the brain remains compatible with—and indeed enhanced by—the system of Freudian thought. Twenty years ago, the general consensus would have been that Freud was as obsolete as phrenology, at least where serious, peer-reviewed, empirical neuroscience was concerned. Today, some of the brightest minds in the field—brain scientists like Jaak Panksepp and Antonio Damasio, neuropsychologist Mark Solms—advocate the building of bridges between Freud's mythological terrain and the new world being mapped by fMRIs and PET scans. The Nobel laureate neuroscientist Eric Kandel published a series of widely discussed articles in the late nineties, outlining the ways in which psychiatry, and psychoanalysis in particular, could be connected to the increasingly rich field of cognitive neuroscience.

Somewhere in this unlikely alliance are the building blocks for the new popular model of the psyche. But to understand what this model will look like, we must first go back to the roots. Here is Freud writing in 1917, by many accounts at the top of his game, as the Great War wound to its close:

Almost all the energy with which the apparatus is filled arises from its innate instinctual impulses. But these are not all allowed to reach the same phases of development. In the course of things it happens again and again that individual instincts or parts of instincts turn out to be incompatible in their aims or demands with the remaining ones, which are able to combine into the inclusive unity of the ego. The former are then split off from this unity by the process of repression, held back at lower levels of psychical development and cut off, to begin with, from the possibility of satisfaction. If they succeed subsequently, as can so easily happen with repressed sexual instincts, in struggling through, by roundabout paths, to a direct or to a substitutive satisfaction, that event, which would in other cases have been an opportunity for pleasure, is felt by the ego as unpleasure. As a consequence of the old conflict which ended in repression, a new breach has occurred in the pleasure principle at the very time when certain instincts were endeavoring, in accordance with the principle, to obtain fresh pleasure.

The apparatus in question, of course, is the human psyche, though it might as well be a steam engine, given Freud's emphasis on its surging, shifting energy. Like almost all his writing, this is a complex, combinatorial language, filled with negations of negations and participatory metaphors. For all its complications, though, I think this passage does an admirable job of conveying both the insights and the blind spots of the Freudian model, at least when viewed through the lens of modern neuroscience. To be sure, the passage does not offer a comprehensive survey of Freud's theory of mind. Parts of it conflict with writings from other stages of his career. This is one of the great problems—and great charms—of reading Freud: he changed his mind at several key points in his intellectual life. *Beyond the Pleasure Principle,* from which this passage is taken, marked just such a turning point. Freud had constructed an entire dynamic model of the psyche with the drive for pleasure as its central piston, and here were these veterans from the Somme endlessly revisiting the horrors of battle in their dreams. The pleasure principle couldn't account for that behavior on its own—it was as though Freud had built a theory around the primacy of the sweet tooth and then found a whole cadre of people who habitually dined on mud and sea salt. Hence this strange notion of "unpleasure" that arises at the end of the passage.

The path of logic Freud follows in developing this idea of "unpleasure" can help us understand how the Freudian model might fare in the age of neuroscience. But to follow that path we have to read more slowly.

Freud begins on a strong note: the energy that fills the apparatus stems from its "innate instinctual impulses." This is the modular theory of the brain, with the explicitly Darwinian framework stripped off, or dampened down. A team of instincts is literally driving the organism, supplying its fuel—or, more exactly, keeping a foot on the accelerator. Of course, instincts and drives were not exactly news even then—Nietzsche had formulated his "will to power" decades before, following Schopenhauer's lead before him. What Freud added to the mix arrives in the next sentence:

In the course of things it happens again and again that individual instincts or parts of instincts turn out to be incompatible in their aims or demands with the remaining ones, which are able to combine into the inclusive unity of the ego.

This is Freud's Copernican idea, the one that fundamentally stood the world—or in Freud's case, the self—on its head. It's as long-decay as they come; a hundred years later, we're still feeling its reverberations. Freud's proposition here is not just that we're com-posed of instinctual drives—it's that those drives are more often than not in conflict with one another, which makes the conscious self less of a controlling subject and more of a battleground. A cer-tain set of drives "combine into the inclusive unity of the ego," where they are experienced as a kind of free will, the self acting on his or her desires for seemingly rational or intuitive reasons. They lose their status as distinct drives and just become a transparent part of the self, its "inclusive unity."

But Freud's breakthrough idea wasn't only of the self as a melee of competing drives. He ventured further: drives that lost out in this competition, that failed to integrate themselves into the ego, didn't just disappear. They hung around like a squad of sore losers, hankering for a rematch. The self, therefore, was not just the sum total of the ego's inclusive unity, the sum of all the drives that made the cut. The neglected ones continued to maintain a presence, even if they operated below conscious awareness. But if those failed drives didn't disappear, where did they go? What happens to a drive deferred?

This was truly the $64,000 question for Freud. Up to this point, his model of the psyche was totally compatible with the modern portrait of the brain's inner life: a host of distinct modules compet-ing for control of the organism, each driven by its own priorities. Depending on the situation, some of those modules would come to influence the brain's executive branch—when our libidinal instincts compel us toward a sexual partner, or our fear response makes us freeze in sudden alarm. When these drives prompt you to act, it doesn't feel like some alien force co-opting your brain; it feels like *you*, like you experiencing sexual attraction or fear. The difference between Freud and modern brain science on this point is partly the number of modules and partly the nature of their interaction. Freud imagined the psyche as a battleground with only a handful of warring forces, most of them sexual in nature. The modular theory, on the other hand, assumes the existence of dozens of specialized tools usually working in an integrated way: face recognition devices, object-naming devices, danger-detection devices, and so on. Sexual instincts are part of that mix, but only a part.

When one of those modules comes to dominate your attention, what happens to the others? You're having a pleasant conversation with your wife, engaged in nuanced mindreading exchange, your brain locked in on her vocal intonations and her subtle shrugs and grimaces and half-smiles. And then you hear, in the background, the rising sound of wind whistling through your apartment windows, and your amygdala initiates the fear response, even as your wife continues talking. Your working consciousness is no longer filled with the subtlety of your wife's speech; it's filled with thoughts about your physical proximity to the window, and your memories of the day it blew in. You're hearing your wife as she continues to speak, but you're not really *listening*. Has your mindreading module been switched off? Or is it continuing to plug away at its task, below the radar of your consciousness? If so, is it somehow frustrated by this demotion?

Freud's answer to these questions was grounded in his concept of repression. Drives that didn't find a way into the inclusive unity of the ego were repressed, blocked from conscious awareness, and left to fester with potentially dangerous consequences:

[They] are then split off from this unity by the process of repression, held back at lower levels of psychical development and cut off, to begin with, from the possibility of satisfaction.

In Freud's model, a fulfilled drive that has found its way to the ego floods the self with feelings of pleasure. Drives that are repressed are denied the possibility of creating that pleasure, and thus seek other methods to reach their goal. It's fair to say that the primary work of psychoanalysis—in the doctor-patient encounter—lies in figuring out exactly what those methods are.

> If [the repressed drives] succeed subsequently, as can so easily happen with repressed sexual instincts, in struggling through, by roundabout paths, to a direct or to a substitutive satisfaction, that event, which would in other cases have been an opportunity for pleasure, is felt by the ego as unpleasure.

Repression does not result in a drive dissipating into nothingness. Instead, it creates a kind of potential energy, confined to the unconscious but seeking ways to escape. Think of those drives as a kind of compressed gas trapped in a small space. The gas "seeks" to escape its confinement through cracks in the walls, or openings beneath the door, as in the revealing emissions of Freudian slips and dream imagery. Build up enough pressure in the chamber, though, and the whole thing explodes—into uncontrolled hysteria, anxiety, madness.

The analyst, in this analogy, is like the guy you call from the gas company when you smell a leak in the basement. He arrives with an array of sensors exquisitely calibrated to detect the source of the leak, and a lifetime of experience hunting down leaks in other houses. You call him up with a vague sense that something's wrong, and before long he's shown you exactly the faulty joint that enabled the gas to escape in the first place. Your repressed drives are trying to claw their way into actualization, and sometimes that entails disguising themselves in baroque dream imagery or uncontrollable hand-washing. Whatever the escape route, the doctor is there to chart its path for you.

The analogy breaks down when you get to the question of actually *fixing* the leak. In the psychoanalytic model, simply talking about the faulty joint—bringing it into the light of consciousness—makes it go away. If you can understand the circumstances in which your drives became repressed, they will no longer plague you. Before analysis, Freud writes, the patient is "obliged to *repeat* the repressed material as a contemporary experience instead of, as the physician would like to see, *remembering* it as something belonging to the past." Shedding light on repressed drives turns out to be much like shedding light on vampires: get them in the noonday sun and they disintegrate.

So this is the broad-strokes rendition of the Freudian model: a battleground of competing drives, some of which lose out and are subsequently pushed underground, where they plot their escape through circuitous means. The talking cure attempts to unearth those drives, thus lessening their hold over the psyche. You don't need to be in analysis to be influenced by this model—most of today's pop psychology works within its framework, even if it's not explicitly labeled as Freudian.

How does this model fare in the light of modern brain science? Some of its insights are as valuable as they were a hundred years ago; some categories need to be translated into a language based more closely on the brain's physiology, as Freud no doubt would have been thrilled to see. And some of the governing metaphors should probably be replaced altogether.

Which of Freud's core concepts remain relevant? Two in particular: the idea of the divided self and that of unconscious processes. You are the sum of your modules, but your conscious sense of self is only a part of that system. Beneath the brain's executive branch, matters are more unruly: a host of subsystems dedicated to registering incoming stimuli, interpreting that data, and making emotional value judgments about its nature, connecting these new developments with past memories, maintaining your body's homeostatic balance. At any given moment, the executive branch may be actively focused on input from a select few of these subsystems.

As I write these words, my attention is divided roughly between two primary actions: thinking about the words as they are generated in my head and materialize on the computer screen, and half listening to familiar songs playing in the background. I am dimly aware of the tactile feeling of my fingers touching the keyboard, though the process of locating the proper keys has become so automated by now that it has dropped below the level of my consciousness. I also have a vague background sense of mood—a bright midmorning working alertness, slightly caffeine enhanced. Those parts of my brain are, in Freud's language, bound up in my ego's inclusive unity. But beneath them, a whole assembly line of mental activity continues to churn away: my amygdala scouring its low-road sketch of incoming stimuli for potential threats; my brain stem regulating my breathing and heart rate and blood sugar levels. Other modules have been worn down out of conscious awareness through regular use: the motor control regions that help my fingers dance across the keyboard with such ease, the language skills that let me type most words without even thinking about how they're spelled. Those specific modules in my brain are processing this knowledge, but I don't need to *think* about that knowledge to make use of it; it seems to come naturally, even though typing and spelling are hardly innate skills. And then there's everything I'm ignoring: the sound of traffic out on the street, the temperature of the air against my skin, the bright yellow color of the wall in front of me. My sensory cortex is processing this input from the outside world, but because I'm focused on the screen and the music, on some level I don't perceive it. If those sirens started to increase in regularity, though, or the temperature suddenly spiked, an internal alarm would go off in my head, bringing the appropriate sensory module into my foreground consciousness.

So even at your most focused moments—even, in Virginia Woolf's words, at your most "pointed, dartlike, definite"—you're still a self divided, the sum of your various modules, some of them rising into the ego's unity, some of them operating behind the curtain. The neurological reality of that curtain—and the hidden world living behind it—endorses one of Freud's most controversial and contested propositions: that our lives are shaped by unconscious mental activity. Every minute of every day, we are shaped by mental calculations that don't bother to report directly to the executive branch. Not only are we divided selves, but some of the divisions don't even show up on our internal radar. The idea of the unconscious was a radical idea in the 1890s when Freud first formulated it, and in many ways it's as radical as ever in the first years of the twenty-first century. Decades of empirical research have endorsed the underlying principle again and again. You can perceive these subsystems through various routes: taking drugs, or doing intense meditation, or spending time with tests or illusions designed to tease out the modular nature of the mind. Or you can follow the path Freud chose: you can just observe carefully.

To be sure, the actual unconscious doesn't quite look like the one Freud imagined. It is not seething with incest fantasies suppressed by the restrictions of civilized society. (With incest, Freud had it exactly the wrong way around: the prohibition against sleeping with blood relations originates with our DNA, not our culture.) Most of the time, in fact, the unconscious is concerned with far less

titillating matters than Freud suggested. Another word for unconscious is "automated"—the things you do so well you don't even
notice doing them. It's stepping on the clutch when you want to
change gears, or flipping your middle finger over your thumb while
playing the last three notes of a piano scale. We're unaware of these
decisions or urges not because they threaten our culture-bound ego
or because they're too explosive for the psyche to handle directly.
We're unaware because we have better things to think about. It's
more efficient for the brain to automate processes that get repeated
a lot. The mission-critical ones—don't stop breathing, flinch when
an object looms suddenly overhead—eventually find themselves
encoded in our genes, while we have to learn the more mundane
repetitions via everyday experience: tying shoes, typing words,
swinging a tennis racket.

Memory researchers call this type of unconscious processing
"procedural" memory, as opposed to the "declarative" memories.
Procedural memory is knowing how to ride a bike; declarative
memory is recalling that time you fell off your bike in seventh grade
and broke your wrist. Simple procedural memories are obviously
not all that interesting where psychoanalysis is concerned: it's nice
that you don't have to think consciously every time you shift gears
in a manual transmission car, but that kind of automated behavior
doesn't reveal that much about the depths of your personality. But
as Eric Kandel points out, some kinds of procedural memories
carry a great deal of emotional weight: when your brain, instead of
simply memorizing repetitive tasks, starts executing complex
assessments of situations on your behalf, without making its criteria explicit. When your amygdala records the clear skies on 9/11
and warns you about potential danger months later on a compara-
bly crisp day. When your mindreading tools catch a flicker of
untrustworthiness in someone's eye, even though you have no clear
sense which micromuscular twitch conveyed that information, and

even less sense of why an eye twitch would tell you anything about the veracity of someone's spoken words. These are not so much procedural memories as they are procedural value judgments, judgments passed without any conscious deliberation on your part. You're made aware of the end results of these calculations—*I feel strangely on edge today; I don't trust this guy*—but the underlying rationale stays behind the curtain.

Those emotional signals can reasonably be described as unconscious drives: forces propelling you in a certain direction without making their reasons clear. But are they unconscious because they've been *repressed*? Here the Freudian model begins to strain. Think of the contortions that Freud had to go through to account for the traumatic flashbacks of his war veteran patients. As it turned out, a simpler explanation existed. But to grasp it, you needed to accept two preconditions.

First, the "drives" propelling the psyche are not exclusively in search of sexual pleasure. For obvious Darwinian reasons, sex is important: your genes don't make it to the next generation if you don't figure out a way to mate with a partner. But if you're killed by a wildebeest before you make it to puberty, your genes aren't going very far either. So our brains evolved systems that rewarded us for having sex, but also pushed us in other directions: toward the bonds of friendship and familial attachment, and away from a host of potential threats. And the way the brain pushes us away from things is by creating feelings of unpleasure—stress, anxiety, fear— in our heads.

The second insight you need is that the drives have their own autonomous relationship to incoming stimuli and stored memories, maintained separately from our normal conscious memory formation. Think of the amygdala's low road and its flashbulb memories of traumatic events; think of oxytocin saturating certain faces with warm feelings of pleasure and contentment. When the amygdala remembers a stray detail from a car accident years ago—a detail you've otherwise forgotten—it's not that the detail was repressed, sent into some kind of mental exile. Your amygdala wants to protect you from threats, and one way it does that is by recording as many details as possible any time you experience danger. Your amygdala is capturing these details not because they're too traumatic for you to handle. It's capturing them because, on certain specific occasions, your amygdala has a better memory than you do.

Put these two ideas together—your brain sometimes protects you by releasing unpleasant feelings, a response itself sometimes triggered by memories that your conscious memory has forgotten— and you have a far simpler explanation for why our brains seem compelled to revisit old traumas. It's not a repressed wish short-circuiting the psyche; it's not some kind of suicide fantasy. It's your brain trying to protect you. I could probably have done without the phobia of sunny days that 9/11 embedded in my brain, but that phobia wasn't a sign that my brain was malfunctioning, repressing some dark fantasy that had somehow become attached to the idea of clear weather. Quite the opposite, in fact. My amygdala was working perfectly. It wasn't some kind of repressive censor. It was more like a sentry, keeping watch while my executive branch went about its business.

These emotional procedural memories don't map clearly onto the Freudian model of repression, but they still have an important role to play in therapy. Building on the work of the psychologist Daniel Stern, Kandel argues that one of the primary goals of therapy may well be the solidification of new procedural memories: replacing harmful "gut" reactions—phobias, emotional estrangement—with more positive ones. In the case of trauma, for instance, you're training your amygdala to resist triggering an alarm at the sight of a snake or an approaching windstorm. The fact that you don't have direct, conscious control over these procedural memories doesn't mean that they've been repressed. They've simply been automated.

You need a theory of repression if you think that the brain's single driving goal is to have sex as often as possible, with as many people as possible (including your mother!). If that's the model, then you have to account for why people spend so much of their time *not* having sex. That's where repression comes in—to keep those drives from being fulfilled. But if you think of the brain as being filled with a much more diverse collection of innate drives—for friendship, social status, safety, aesthetic beauty, novelty, not to mention a built-in incest taboo—then there's not nearly the same need for a repressive model. People spend large parts of their lives not having sex for a simple reason: they've got other needs to satisfy. This is where Freud underestimated just how divided the psyche truly is. The ego is not torn between two masters, contrary to what he famously wrote, strung out between the competing drives of the superego and the id. Modern neuroscience has complicated that power struggle almost beyond recognition. Even the sanest among us have so many voices in our heads, all of them competing for attention, that it's a miracle we ever get anything done.

The pandemonium in our brains returns us to the earlier question about repressed wishes. What happens to a voice that goes unheard? Does it come back to haunt us, as Freud imagined? This is one of those places where Freud's metaphoric scaffolding ended up misleading him. If you think of the brain as a kind of steam engine, filled with energy that seeks release, then repressed drives are either stored somewhere in the brain or they discover indirect outlets to liberate themselves. It's the first law of thermodynamics applied to the mind: the conservation of psychic energy.

But all that changes if you use another metaphor: the brain as Darwinian ecosystem, instead of steam engine. This is a metaphor proposed by the brilliant neuroscientist Gerald Edelman, who won a Nobel Prize for his research into the immune system in the early '70s, and who has subsequently devoted much of his research to the brain. Edelman believes that the internal mechanisms of both the brain and the immune system run miniversions of natural selection. Think of those modules in your brain as species competing for precious resources—in some cases, they're competing for control of the entire organism; in others, they're competing for your attention. Instead of struggling to pass their genes on to the next generation, they're struggling to pass their message on to other groups of neurons, including groups that shape your conscious sense of self.

Picture yourself walking down a crowded urban street. As people pass you by, your face-recognition module scans their features, looking for a match: a friend's visage, or a celebrity's, or a long-lost high school classmate's. Your olfactory centers report the smell of bread being pulled out of an oven as you pass a bakery, which in turn lights up your hunger centers. A sudden horn blast from a truck sends a flash through the low road to the amygdala, which sends a small alarm out suggesting that something may be awry. As you walk, your brain is filled with these internal voices, alongside dozens of others, all competing for your attention. At any given moment, a few of them are selected, while most go unheeded. The truck horn might cause a minor flinch, and you might think for a second that you've just seen your college roommate pass by, but you might be so engrossed in thought that you don't notice the bakery smell or the growl in your stomach.

In this psychic ecosystem, as in real-world ecosystems, failures abound. This is good news. You want all the modules in your head doing their best to persuade your executive branch to pay attention; you want your blood sugar monitored and your memories recalled. But you want those acts of persuasion to fail most of the time so that you can focus minute by minute on the important issues, the ones that have made the most persuasive case. In the Freudian steam engine, a repressed drive eventually finds a path to fulfill-ment, even if it damages the individual along the way. Failure is not an option. But in the Darwinian model, failures are a sign of success.

Does this mean that Freud simply hallucinated the baroque lan-guage of dream symbolism? If unconscious drives could disappear without causing further harm, if they no longer needed to find alternative routes to express themselves, then why were dreams so loaded with emotionally charged symbols?

In fact, Freud's insight here remains a valuable one, though once again, you have to fiddle with the categories to make it work. Your dreams, or passing thoughts, or slips of the tongue can sometimes contain unintended—but nonetheless revealing—connections to emotionally fraught memories or desires (revealing precisely because they're unintended). But those revelations don't come about because the unconscious needs to speak in code to avoid the superego's impe-rious censor. They come about, first and foremost, because the brain is an associative network in which thoughts—the memory of a fifth-grade field trip, the concept of "transference," the color red—are rep-resented by groups of neurons distributed throughout your brain that fire in sync with each other. Certain thoughts have more neurons in common than others. Neurons that fire together wire together. A cigar may still be a cigar, but its shape might indeed trigger some of the same low-level object-shape-recognition neurons that the sight of a penis does. Which means that every now and then, thinking about a penis might trigger the image of a cigar, and vice versa. If our brains weren't wired this way, we'd be incapable of poetry, as well as most abstract learning.

These connections are not your unconscious speaking in code. They're much closer to free-associating. These revelations aren't the work of some brilliant cryptographer trying to get a message to the frontlines without enemy detection. They're more like echoes, reverberations. One neuronal group fires, and a host of others join in the chorus.

So why do so many of our free associations gravitate toward emotionally fraught topics? The answer should be obvious by now. Our emotions and our memory are locked in a deep embrace: memories experienced under the influence of strong emotion are more easily recalled. Emotions affect the way we feel, but they also affect the way we remember. On the whole, we're more likely to remember emotionally charged memories than we are emotionally neutral ones. This tips the scale in the free-association game—in our dreams and our waking states—toward the more powerful thoughts: thoughts of sexual pleasure or frustration, sudden fear, social camaraderie, parental love and parental anxiety. The big issues, in other words. Associative networks like to riff, but they also have a fondness for the old standards.

Which brings us to the question of cure. When we revisit these emotionally charged memories, prodded along by our analyst's questions or by our own introspection, does this exposure lessen their hold over us? Does it, as Freud described, help us move from repeating the past to remembering it? The answer revolves around whether our emotional systems are activated again in the process of conjuring up the memory. If the emotions come flooding back to you when you think of some past event—if you feel the terror swell up inside you, or you convulse with sadness—then you're only increasing the emotional weight of the memory. Even if, in the course of retriggering it, you learn something about why the emotional memory of the event in question is so strong. It can seem cathartic to us to relive powerful events in all their emotional intensity, but because of the way the brain's emotional and memory systems interact, reliving events only makes them stronger. With some traumatic events, you may in fact be better off simply forgetting.

But what about events that we *can't* put out of our mind, either because they return compulsively to our thoughts or because our daily routines force us to confront them, like my memory of our window blowing in? This is where the talking cure can help, for reasons that brain science can readily explain. The way to progress from repeating to remembering is to wire your brain so that re-creating the event in your head no longer unleashes an emotional response. On some level, we're back to the domain of the behaviorists here, and the tone-shock experiments. As discussed earlier, if you hear a tone, and then experience a shock, you'll develop a fear of tones. If you hear wind whistling through a window, and then the window comes crashing into the room, you'll develop an anxiety about the sound of wind. The way out of these damaging associations is to make new associations. I still get a little on edge every time I hear the wind rise up outside our apartment, but my anxiety levels have eased markedly over the past few years, because I've heard that sound hundreds of times without the window shattering. Ever so slowly, the sound is becoming associated in my brain with safety— with windows miraculously staying in their frames. Something similar can happen on the therapist's couch: you re-create the traumatic memory in a safe environment, and by doing so you rewire your neural associations. (Scientists have a lovely Darwinian term for the passing of that older association: "extinction.") In therapy, your childhood traumas slowly become associated with a relaxed posture, pleasant decor, and a comforting presence in the room with you. It's not so much *understanding* the source of the anxiety that helps you; it's reliving the trauma without something negative happening again, thus forging a new association in your head and dampening the original emotional response. We're back to the idea of reconsolidation: when we relive a memory, we make a new memory in the process, with new connections. All of our remembered pasts are transformed by the present.

It's worth pointing out that what applies to negative emotional memories applies to positive ones as well. If you want to lessen the power of a traumatic memory, don't endlessly revisit it without actively trying to forge new associations in the process. If you simply conjure up the emotional response again and again, you'll just dig yourself a deeper hole. Positive emotional memories—career triumphs, sexual intimacy, social bonding—work the same way, but of course we generally want positive memories to have *more* sway over our lives, not less.

Think of this as the neurochemical argument for savoring. If you find yourself in a car accident, do whatever you can (including perhaps taking beta-blockers) over the next few weeks to avoid reliving the event and triggering the fight-or-flight response all over again. But if you win an award, or have a great conversation with an old friend, or publish the novel you've been working on for years—if something happens that makes you feel unusually happy—take time out over the next few weeks to savor that experience, to remind yourself of how it made you feel. By doing this, you create a kind of virtuous feedback loop in your brain: you deepen the emotional weight of the memory, and thus make it more likely to influence your thoughts and actions down the line.

There's a classic stereotype of the chronic overachiever who's never satisfied with his latest success, and who's always striving for the next one. But I suspect that most successful people genuinely enjoy success, and seek out more of it because they like the way successes make them feel. If you're the kind of person who doesn't like to dwell on your accomplishments, get over it. If it's good news, by all means dwell.

We've kept the core insights of the Freudian model: the divided self and the unconscious. But the guiding metaphors have changed: the brain is more Charles Darwin than James Watt, more ecosystem than steam engine. Our unconscious thoughts are not repressed by an austere censor, and many feelings of unpleasure that they trigger are signs of a functional psyche, not a dysfunctional one. The brain is more likely to free-associate than speak in code, though any free-associating sojourn is likely to lead back to emotionally charged memories. And where those charged memories are concerned, the brain needs to do more than just understand their origins to shake them off—it needs to make new emotional associations.

If this is Freud with new metaphors, then what about the proper names? Freud divided the psyche into three primary subselves: id, ego, and superego (roughly parallel to unconscious, conscious, and preconscious). If we're attempting a neurologically correct draft of the Freudian script, who are the new lead characters?

The pop psychology version closest to Freud is arguably the left-brain, right-brain split, which is real enough but probably doesn't deserve to be at center stage. There is certainly an important division of labor between the left and right sides of your brain, most significantly the seat of language on the left. But there is also a great deal of redundancy and shared function between the two hemispheres, not to mention the extensive communication channel that connects them via the corpus callosum. The two sides might be likened to two different facets of the ego: one slightly more gifted verbally, the other better with spatial logic.

The closest neuroanatomical equivalent to Freud's id/ego/superego is the "triune brain," proposed half a century ago by Paul Maclean. Maclean's vision of the brain's organization is both an evolutionary story and a topographic one. To use a metaphor that Freud himself employed in *Civilization and Its Discontents*, our brains are a kind of archaeological dig site, with a series of settlements stacked one on top of the other. The deeper you dig, the farther you go back in time. At the deepest level lies the reptilian brain, also known as the brain stem, controlling the body's basic metabolic functions, like heart rate and breathing. The brain stem is all primitive instinct and repetition, incapable of emotional complexity or anything resembling genuine thought.

The second layer in the triune brain is known alternately as the paleo-mammalian brain and, more famously, the limbic system. This is the seat of emotion and memory, comprising chiefly the amygdala, the hippocampus, and the hypothalamus. Our primary emotions—love and fear, sadness and joy—emerge from this region, coloring incoming stimuli with the emotional valences we've associated with past events stored in the hippocampus or the amygdala. We share this architecture with most mammals, which is one reason we're capable of forming powerful social bonds with fellow mammals like dogs or horses—not to mention our closest relatives, chimps. Dogs and cats are vastly more popular as pets than lizards and snakes precisely because they seem to have a much more dynamic emotional repertoire. When we sense emotional complexity in other mammals, we're detecting the existence of the limbic system operating in their brains.

Stacked on top of the brain stem and the limbic system is the neocortex, the two hemispheres of which spread across the surface of the brain like a foam insert in a bike helmet. This is the most distinctly human component of the brain's architecture. Only our primate cousins have anything close to it in size, though extremely small versions of the cortex have been discovered in the brains of rats and other mammals. When we alter our immediate actions because of long-term interests, when we communicate in complex sentences, when we engage in abstract thought—indeed, when we display most of the hallmarks of human intelligence, we're most often using our neocortex.

Maclean's model has fared remarkably well over the past fifty years. The basic evolutionary story—the movement from brain stem to neocortex in the progression from reptiles through mammals to primates—is now widely accepted. Of the triune brain's three primary actors, the limbic system remains the most controversial. Some scientists agree with the general description of its function but don't believe that it operates as a coherent system. Certainly there is much interplay between the neocortical capacity for reason and the limbic system's emotional judgments. As Antonio Damasio has amply demonstrated over the years, people with damage to their emotional centers are often incapable of rational decision-making, because our emotional centers provide quick, instinctive responses to situations that a purely rational brain might have to cogitate over for hours. Memory, as well, complicates the limbic system model, since memories are so crucial to both emotional and rational processing. Indeed, one of the first challenges to the limbic theory came from studies of patients with damage to their hippocampus, which resulted in significant cognitive problems because of the hippocampus's role in long-term memory formation.

Like any complicated archaeological site, Maclean's historical map of the brain has its points of contestation. If our brains are like three separate towns stacked on top of one another, some buildings may well turn out to have been used by residents from two different eras, and the demarcation between the ancient settlements and the newer ones may turn out to be blurrier than we originally thought. But the general progression from brain stem to limbic system to neocortex—as E. O. Wilson puts it, from heartbeat to heartstrings to heartless—is certainly a more accurate assessment of the psyche's inner divisions than the old mythos of id, ego, and superego.

Within this newer topography, there are a handful of critical hubs, most of which we've explored over the preceding pages: the amygdala, the hippocampus, the "executive brain" in the frontal regions of the neocortex. But just as important for a popular understanding of the brain's inner life are the molecules of emotion and affect: oxytocin, cortisol, serotonin, and so on. These chemicals constitute the raw material of the brain's value system. They are in a sense the closest equivalent to Freud's idea of "energy" filling the apparatus of the mind. If we're going to rewrite the language of selfhood along the lines suggested by modern brain science, these agents—and their effects—have to be part of our vocabulary. Learning to recognize their presence should be a touchstone of the examined life. Serotonin's rejection-insensitivity and social confidence; dopamine's exploratory push, its seeking without pleasure; cortisol's frayed edge; the endorphins' oceanic bliss; oxytocin's drive to make emotional bonds; adrenaline's sudden lift. These are the humors of the modern world—the drugs in your inner medicine cabinet, the chemicals your brain relies on to push you toward certain objectives and away from others.

All of us blessed with normally functioning neurological equipment share this chemistry. But the question remains: which drugs get doled out when? Our personalities—the entities that make us both unique and predictable as individuals—emerge out of these patterns of chemical release. Part of what makes me *me* is that my brain has been wired to release adrenaline when I get a good laugh from someone, and cortisol on sunny days, and endorphins when I stand next to the crib and watch our son sleeping. Whether that wiring comes courtesy of my genes or my lived experience, or via some combination package, is not necessarily relevant. What matters are the incoming stimuli and the pattern of activity that they spark: your brain taking in a certain configuration of sensory data from the outside world (or from your imagination or your memory banks) and triggering a neurochemical reaction in your head.

Pattern recognition instead of code breaking—this may be the simplest way to describe the difference between the twenty-first-century Freud and the original. The two approaches can be readily blurred together. After all, you need pattern recognition tools to break a code. But breaking a code involves a further step—translating the encoded message back into its original form. In reality, those patterns in my own head don't conceal a secret meaning that analysis can unearth after intense scrutiny; they don't have symbolic depth. They don't speak in ciphers. My fear of wind doesn't represent some submerged anxiety from my childhood; it's the imprint of a pattern that my amygdala first detected on that June afternoon: the wind howls, and then shattered glass is flying everywhere. In fact, there are two patterns here—the initial chain of events and then the sensory-neurochemical chain (hear wind, initiate fear response) that repeated so many times in my head that it became unshakable. Knowing something about my brain's inner life helped me see that pattern more clearly. But seeing the pattern clearly didn't entail discovering some deeper meaning buried like a prize in a box of cereal.

Chances are, learning to recognize these patterns won't make them go away. But if you know something about your mental medicine cabinet, you'll be able to take into account ways in which these chemicals bias your judgment. So when you sit down to balance the checkbook feeling flush with serotonin, you learn to recognize that you're likely to see the glass half full (if not frothing over) under that particular influence, just as a trip to the accountant under a cortisol cloud will likely make you want to stick your head in the oven. Neither perspective grants what you would call an accurate assessment of the real world; both contain a distinct interpretative slant. And you can't just wish these feelings away when you're under their spell, either through force of will or by decoding their secret meaning. What you *can* do is recognize the pattern of chemical release, and if your response to the situation doesn't seem appropriate, you can discount the drug's effects.

* * *

So this is your brain, in all its multiplicity. You are part reptile, part mammal, part primate, part homo sapiens. You are a twitchy amygdala; you are a dopamine fiend; you are under the spell of oxytocin. You are an unthinkably complex series of connections, of links, spun together by your genes and by your lived experience. You are a walking assembly of patterns and waves, clusters of neurons firing in sync with one another.

When I talk to people about this vision of the mind—people who have not followed recent developments in brain science—most of the time, their response is genuine interest and recognition. They nod their heads a lot, and seem to find an immediate connection to the ideas. But a substantial minority has another response. You can see them flinch ever so slightly as I talk about the brain's subsystems, as though I were describing something viscerally disturbing, more than a little creepy. There's a kind of vertigo induced by this line of thought: you catch a glimpse of your own mind as an electrochemical grid, all those separate modules churning away beneath the surface of your awareness, and the world starts reeling.

Freud's model of the psyche had a comparable effect on its initial audience. During the writing of *Beyond the Pleasure Principle*, Freud spun off a short, cryptic essay called "The Uncanny." The essay ruminated on some of *Beyond the Pleasure Principle's* general themes—repetition compulsions, death drives—but looked ultimately at the question of why we get spooked by strange coincidences and superstition. When we find something uncanny—the same numeral recurring several times in different contexts over the course of a day, catching yourself in the mirror and not recognizing yourself for a few seconds—where does this feeling come from? In the essay, Freud remarks: "I should not be surprised to hear that psychoanalysis, which is concerned with laying bare these hidden forces, has itself become uncanny to many people for that very reason. In one case, after I had succeeded—though none too rapidly—

in effecting a cure in a girl who had been an invalid for many years, I myself heard this view expressed by the patient's mother long after her recovery."

I think there is something valuable in this uncanny reaction. In fact, I have tried to cultivate it in myself. Most of the time, I suspect I'm walking around with one of two models of my mind in the foreground of my attention: either the intuitive unified self or the modular neural brain. I can now switch back and forth between these two images with relative ease. But every now and then, I manage to hold both images in my head at the same time: I'm me *and* I'm a big lump of neurons. That's when I feel a flash of the uncanny. It's an honest feeling, the mind sensing the fundamental contradiction it embodies—that you are both one and many at the same time.

There's another response to these ideas, one that I have less patience for. And that is the idea that there is something demystifying about this perspective, something soul-depleting or artless. The poets and philosophers are supposed to explain our mental life, not the fMRI machines. By turning ourselves into a squad of walking neural nets, we're "unweaving the rainbow," to borrow a Keats phrase courtesy of Richard Dawkins. We're taking something magical and reducing it to a crude piece of machinery.

I think this reaction is wrong for two reasons. First, because there is plenty of magic to go around in both the technology and the insights of modern brain science. Being able to peer into your brain and see those microscopic patterns of blood flow and electrical activity, to see yourself thinking on the level of actual neurons—that vision is truly indistinguishable from magic. And there is no conjurer's trick in nature more profound than the human brain's capacity to create a sense of unified selfhood out of dozens of competing neural systems. The more you learn about how the brain actually works, the more magical the apparatus seems. The more you learn about the brain, the more you understand how exquisitely crafted it is to record the unique contours of your own life in those unthinkably interconnected neurons and their firing patterns. Brains come with a common architecture, and it can be thrilling to explore those commonalities—with our fellow humans, of course, and also the primates and reptiles that share some of that architecture. But part of that architecture evolved to record and amplify individual differences, the imprints of our personal trajectories through the world.

When I watch my son sleeping, and feel the contented shiver of opioid release as I gaze into his crib, part of the wonder of that experience is its connection with the history of mammals and their evolved child-rearing systems, the miracle of the tending instinct. But another part of the wonder lies in the details, in the knowledge that this precise pattern of neurons firing in my visual cortex—the pattern that corresponds to the soft edges of his face, half illuminated by the night-light—belongs to me and me alone. Knowing something about your brain chemistry at such a moment connects you both to the individual neuronal assemblage in your brain that creates the image of your child and to the evolutionary history of feeling, the history of all your ancestors and their parental emotions. If there is not grandeur in this vision of life, to use Darwin's famous phrase, then grandeur has become meaningless. It doesn't make me love my son any less, standing there in the dark at the side of the crib, knowing something more about where love comes from.

There is a second objection to the demystification argument, and it revolves around the idea of "reductionism." When people complain about scientific or biological attempts to explain human behavior, what they're often saying is that science "reduces" human complexity to biological component parts, and in that reduction, some essence is lost. The rainbow is just refracted light, the brain just a box of competing modules. Of course, anyone who has spent any time actually reading the scientific literature on the brain knows that the current model of how the brain works is an immensely complex one, hardly a crude simplification. It is vastly more complicated and multilayered as a theory than Freud's theory of mind was, more elaborate than Shakespeare's or Aristotle's. Actual individual brains are of course more complicated than any theory that describes them, and so in building a model of brain function, there is a necessary reductive step in moving from object to model. But that is true of *any* attempt to explain the mind's behavior, whether it takes the form of a sonnet, a philosophical discourse, or a peer-reviewed paper in the *New England Journal of Medicine*.

In a sense, the debate about reductionism here intersects with the critique of "biological determinism" in the debate about evolutionary psychology and the nature-nurture divide. Some believe that any attempt to talk about the human psyche using the tools of science is an encroachment on terrain that properly belongs to the humanities: the men in white lab coats infiltrating the ranks of poets and historians and sociologists. Because the human mind creates culture, it should be up to the cultural producers, not the scientists, to explore the inner life of the mind. But this critique is only valid if scientists are proposing to do away with cultural interpretations altogether.

Which they're not. What *has* been proposed—and what this book in its own way has tried to help bring about—is a bridging of the two worlds: of biology and society, nature and nurture, science and humanities. We're back to Henry James here, and his discriminating eye. James and other classic novelists helped us see patterns in our own behavior, in our mental engagements with the world. Brain science can do the same, either by zeroing in on the specific constellation arranged in your own head (via neurofeedback or brain imaging) or simply by teaching you to listen better to your own inner life, to detect the release of certain chemicals or cognitive modules. Understanding the biological workings of our brain can throw into sharper relief the achievements of culture; it can also suggest ways that society could be made better. And just because our mental modules are implicated in political issues—in our capacity for trust, for social connection, for stress and anxiety—that's no reason to hand over our societal reins to the evolutionary psychologists or neuroscientists. To include biological perspectives in a discussion of human society by no means eliminates the validity of other kinds of explanations. What people like E. O. Wilson have proposed is not biological determinism, but rather biological consilience: the connecting of different layers of experience, each with its own distinct vocabulary and expertise, but each also possessing links up and down the chain. Steven Pinker describes it wonderfully:

> Good reductionism (also called hierarchical reductionism) consists not of replacing one field of knowledge with another but of connecting or unifying them. The building blocks used by one field are put under a microscope by another. The black boxes get opened; the promissory notes get cashed. A geographer might explain why the coastline of Africa fits into the coastline of the Americas by saying that the landmasses were once adjacent but sat on different plates, which drifted apart. The question of why the plates move gets passed on to the geologists, who appeal to an upwelling of magma that pushes them apart. As for how the magma got so hot, they call in the physicists to explain the reaction in the Earth's core and mantle. None of the scientists is dispensable. An isolated geographer would have to invoke magic to move the continents, and an isolated physicist could not have predicted the shape of South America.

This consilient approach is not reason for writers to begin every biography with the emergence of multicellular organisms, or to explain the rise of Impressionism starting with the physics of light. If that were the case, you'd have to launch every book with the story of the Big Bang, and you'd never get anywhere. Traditional narratives that keep to a single explanatory layer are wonderfully enlightening in their own right, and the good news is that the bookstores and libraries are stocked amply with them. But those narratives are only part of the story. There is no convincing reason a comprehensive account of the self in society couldn't be built by a consilient chain: neuroscientists explain how the brain's underlying electrochemical networks function; evolutionary psychologists explain how and why those networks create channels of "prepared learning" or instinct; sociologists explain what happens when those channels come together in large groups of individual minds; political theorists and moral leaders explore the best ways to structure society to reconcile those patterns of group behavior with individual needs; historians tell us how all these various layers have ended up clashing with history's roulette wheel.

Including a few layers of biological knowledge in this chain doesn't hijack the process; it doesn't turn us into slaves to our neurons or our DNA. In fact, the addition might well make our cultural systems more effective by illuminating useful avenues to explore, and suggesting areas where our brain's faculties may create too much resistance. The more we understand our nature, the better we'll be at nurturing.

The brain is the beginning of human culture, which makes culture an outgrowth of the brain's biology, like a bloom on a vine: more beautiful than its support system, to be sure, but shaped by that system nonetheless. To grasp the true story of our lives in its entirety, we have to move beyond the bloom, past the poetry and the philosophy and the Henry James novels, down to the level of our brains in themselves as they really are. That this is even possible is one of the great miracles of our time. The mind is now open to us in ways that exceed the wildest dreams of poets and philosophers. Why not peer inside?

THE HUMAN BRAIN

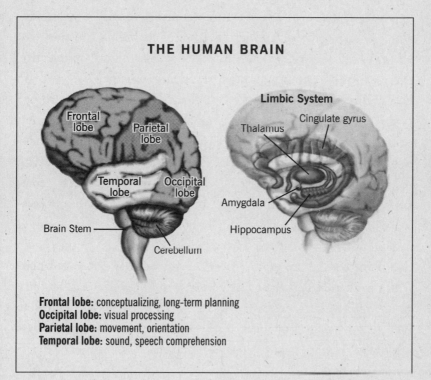

Frontal lobe: conceptualizing, long-term planning
Occipital lobe: visual processing
Parietal lobe: movement, orientation
Temporal lobe: sound, speech comprehension

PREFACE

1. "The idea for this book": A word about the title. There is a widely recognized distinction in the brain sciences and in psychology between "mind" and "brain." The former refers to the experiences we have direct access to—drives, sensations, fears, memories—while the brain is everything behind the curtain: neurons and neurotransmitters and synapses. One way of thinking about the relationship between the two—using language from my last book—is to consider the mind an emergent property of the brain: a whole that is somehow greater than the sum of its parts. The connection between the two levels continues to have much mystery to it, but a number of solid bridges have been built in recent years, whether we're talking about fear, or memory, or attention, or even love. I've tried to keep this story to the most solid structures, though I've also included more speculative research where it seemed appropriate. I haven't hesitated to move from the level of mind to the level of brain and then back again—in a way, the whole point of the book is that this sort of level-jumping has great potential for personal insight. So the title is not intended to suggest a focus on mind over brain, but rather an opening up of the mind that lets us see the brain's activity in new light.

3. "feeling of mirth": Kurzweil, 149.

4. "brains are like fingerprints": "Monkeys trained to use a certain finger to solve a behavioral task gradually exhibit larger areas of cortical representation for that finger. This may also help explain how an aspiring pianist gradually becomes a skilled artist, and it has been shown that right-

handed guitarists have richer cortical representations of that hand within their left hemispheres. But such plasticity does not tell us why, across different individuals and species, the representations of fingers are found in essentially the same relative locations within their brains—a brain area that in humans is situated just beneath the temples near the tip of the ears. The rest of the body is also represented systematically (and upside down, with one's rear pointing up and the head down, as if one were getting a spanking) on nearby tissue of the precentral and postcentral gyri. The cortical areas for bodily representations are encoded, in some presently unknown way, within the same genes of all mammals." Panksepp, 16.

6. "more like an orchestra": The metaphor is borrowed from Jim Robbins's excellent book on the history of the neurofeedback movement, *A Symphony in the Brain.*

6. "neurochemical release": "The collection of neural patterns which constitute the substrate of a feeling arise in two classes of biological changes: changes related to body state and changes related to cognitive state. The changes related to body state can be achieved by two mechanisms. One mechanism involves what I call the 'body loop.' It uses both humoral signals (chemical messages conveyed via the bloodstream) and neural signals (electrochemical messages conveyed via nerve pathways). As a result of both types of signals, the body landscape changes and is subsequently represented in somatosensory structures of the central nervous system, from the brain stem on up. . . . The changes related to cognitive state are generated when the process of emotion leads to the secretion of certain chemical substances in nuclei of the basal forebrain, hypothalamus, and brain stem, and to the subsequent delivery of those substances to several other brain regions. When these nuclei release neuromodulators in the cerebral cortex, thalamus, and basal ganglia, they cause a host of significant alterations of brain function." Damasio, 1998, 281.

6. "specific regions": " . . . the brain induces emotions from a remarkably small number of brain sites. Most of them are located below the cerebral cortex and are known as subcortical. The main subcortical sites are in the brain-stem region, hypothalamus, and basal forebrain. One example is the region known as the periaqueductal gray (PAG), which is a major coordinator of emotional responses. The PAG acts via motor nuclei of the reticular formation and via the nuclei of cranial nerves, such as the nuclei of the vagus nerve. Another important subcortical site is the amygdala. The induction sites in the cerebral cortex, the cortical sites,

include sectors of the anterior cingulate region and of the ventromedial prefrontal region." Damasio, 1998, 60–62.

6. "the emotion you feel": One of the overarching narratives silently guiding this book is the explosion of interest in the neuroscience of emotion— sometimes called "affective neuroscience," as in the science of affect. I explore other topics in the book, of course, but without real research into emotion, the idea of brain science helping you understand *yourself* wouldn't ring true (unless you were Mr. Spock). I think of my argument here as a kind of celebration of science's new inquiry into the emotional brain. The work of Joseph LeDoux, Antonio Damasio, and Jaak Panksepp has been especially instructive to me.

8. "one called 'cortisol'": "Cortisol levels (the amount of hormone in the blood) are high in many acutely depressed adults. Autopsies often show the adrenal gland to be enlarged in adults who have died by suicide. The gland is also enlarged, according to imaging studies, in about a third of depressed patients. Put briefly, elevated, nonsuppressible cortisol levels can be a marker of depression. Elevated cortisol may even account for certain symptoms of depression. The adrenals are far from the brain, and most of what interests researchers is not so much cortisol produced by the adrenals as the substances in the brain that stimulate the adrenals. There is a cascade of such hormones; one brain center stimulates another, and so on down the line until a hormone is released that causes the adrenals to produce and release cortisol. At the top of the cascade is a substance produced in the brain called corticotropin-releasing factor (CRF). Elevated CRF levels can be measured in the brains of rats subjected to stress—and here is where a more homologous model of stress and depression emerges." Kramer, 115–16.

8. "fundamentally different speeds": "Peptides represent a large class of slow-acting modulatory substances found throughout the brain. They are made up of many amino acids, and are larger molecules than simple amino acids like glutamate or GABA. Because peptides are often present in the same axon terminal as glutamate or GABA (but in their own separate storage compartments), they are released with the fast transmitter when an action potential comes down the axon. . . . But peptides bind to distinct postsynaptic receptors and can, as a result, augment or reduce the effect of the fast transmitter with which they are released. However, since peptides are slow to affect the postsynaptic site, and their effects are long-lasting, they tend to have more of an effect on subsequent squirts of fast transmitter. While glutamate and GABA can have slow effects as well as

fast ones, depending on the receptors involved, peptides typically only have slow modulatory actions. They can affect dramatically the ability of a cell to be fired by other inputs, but cannot do so with precise timing." LeDoux, 2002, 57.

8. "the feeling stays alive": Much of that feeling happens in the body, not the brain. "After forming mental images of key aspects in the scenes (the encounter with the long lost friend; the death of a colleague), there is a change in your body state defined by several modifications in different body regions. If you meet an old friend (in your imagination), your heart may race, your skin may flush, the muscles in your face change around the mouth and eyes to design a happy expression, and muscles elsewhere will relax. If you hear of an acquaintance's death, your heart may pound, your mouth dry up, your skin blanch, a section of your gut contract, the muscles in your neck and back tense up while those in your face design a mask of sadness. In either case, there are changes in a number of parameters in the function of viscera (heart, lungs, gut, skin), skeletal muscles (those that are attached to your bones), and endocrine glands (such as the pituitary and adrenals)." Damasio, 1995, 135.

9. "the feeling of what happens": Damasio, 1999.

10. "call these properties 'qualia'": "Qualia, individual to each of us, are recategorizations by higher-order consciousness of value-laden perceptual relations in each sensory modality or their conceptual combinations with each other. We report them crudely to others; they are more directly reportable to ourselves. This set of relationships is usually but not always connected to value. Freedom from time allows the location in time of phenomenal states by a suffering or joyous self. And the presence of appropriate language improves discrimination enormously; skill in wine tasting, for example, may be considered the result of a passion based on qualia that are increasingly refined by language." Edelman, 1992, 136.

11. "on today's consciousness stage": For some stimulating—and sometimes dizzying—explorations of consciousness, see Damasio's *The Feeling of What Happens*, Dennett's *Consciousness Explained*, Taylor's *The Race for Consciousness*, and Penrose's *The Emperor's New Mind*.

13. "written in the same ink": Joseph LeDoux puts this admirably in *Synaptic Self:* "Let's start with a fact: People don't come preassembled, but are glued together by life. And each time one of us is constructed, a different result occurs. One reason for this is that we all start out with different sets of genes; another is that we have different experiences. What's interesting about this formulation is not that nature and nurture both contribute to

who we are, but that they actually speak the same language. They both ultimately achieve their mental and behavioral effects by shaping the synaptic organization of the brain. . . . The particular patterns of synaptic connections in an individual's brain, and the information encoded by these connections, are the keys to who that person is." LeDoux, 2002, 22.

14. "on different drugs": "The human body contains a mechanism that causes the brains of boys and the brains of girls to diverge during development. The Y chromosome triggers the growth of testes in a male fetus, which secrete androgens, the characteristically male hormones (including testosterone). Androgens have lasting effects on the brain during fetal development, in the months after birth, and during puberty, and they have transient effects at other times. Estrogens, the characteristically female sex hormones, also affect the brain throughout life. Receptors for the sex hormones are found in the hypothalamus, the hippocampus, and the amygdala in the limbic system of the brain, as well as in the cerebral cortex." Pinker, 2002, 281.

14. "politics over science": If you find the idea of a biological difference between the brains of men and women troubling, keep these two points in mind. First, we're talking about averages here, not absolutes. Men on average are more prone to violence than women, but any given woman might well be more violent than any given man. Second, and perhaps more important, the tendencies that brain scientists describe are not set in stone; violence isn't a software program that male brains are forced inexorably to run. Most evolutionary psychologists shun the word "instinct" precisely because it implies something too fixed, too inescapable. Instead, they use the phrase "prepared learning." Natural selection doesn't hand down a strict playbook for action—it offers hints and clues instead. We find it easier to learn strategies that are part of our toolbox than we do strategies that weren't adaptive in our ancestral environment. You have to go to school to learn how to read, but no one goes to school to learn how to read facial expressions, although it is an incredibly sophisticated art. On the other hand, whatever we're prepared to learn can be unlearned under the right circumstances. And the fact that we're prepared to learn a certain type of behavior says nothing about the social or political merits of that behavior. Men may be prone to violence, but that doesn't mean as a society we have to accept violence. We overcome our so-called instincts all the time with no political repercussions whatsoever. We fly in planes and work in skyscrapers despite a fear of heights reasonably endowed to us by natural selection. That doesn't make life at thirty thousand feet immoral. It just makes it a

little more difficult to pull off than life on the ground, where natural selection expected us to be.

16. " 'long-decay' test": Peter Kramer's *Listening to Prozac* illustrated precisely this long-decay approach—exploring the ways that drugs that modify serotonin availability in the brain in turn change the way that we think about ourselves as individuals, and as a society. "We may become more aware of our own feelings of confidence or despondency, noting how they respond to our social circumstances—how applause is a tonic for us, how loss devastates. We will no doubt worry over our depressions as once we worried over carcinogens: are they causing covert damage? An unreliable lover enrages us—he is doing not just psychic but physical harm; we assume the two are much the same. Or we see our spouse as a sort of first neurotransmitter in a cascade of chemicals, one who keeps our serotonin levels high. We are keenly aware of our temperament, our psychic scars, our animal nature. Assessing both ourselves and others, we find ourselves attending to strange categories: reactivity, aloneness, risk and stress, spectrum traits, dysthymic and hyperthymic personality. We understand that our reliance on biological categories has run far ahead of evidence, but we are scarcely able to help ourselves." Kramer, 296.

1: MIND SIGHT

22. "this phenomenon as 'mindreading'": For a book-length exploration of this topic, see Baron-Cohen, 1999.

22. "duet for granted": I think of this process as being like the dramatic technique that Bertolt Brecht called the "distanceation effect." Radical theater, in Brecht's vision, was supposed to distance us from our too-familiar social structures, make us see those structures with fresh eyes. In this book we'll do the same thing again and again with aspects of human experience that we've long taken for granted, so much so that we stop noticing them. This is one of several ways in which the objectives of the arts run parallel to the lessons of brain science—in a sense, they both aim to get you outside of your head in order for you to see your head better.

23. "mirror neurons": See V. S. Ramachandran's fascinating overview of mirror neurons and their evolutionary significance, "Mirror Neurons and Imitation Learning as the Driving Force Behind 'the Great Leap Forward' in Human Evolution," archived at www.edge.org/documents/archive/edge69.html.

23. "origins of language": Rizzolatti and Arbib, 1998.

24. "deaf-blind children": Wilson, 153.
24. "University of California at San Francisco psychologist Paul Ekman": Ekman gives a revealing account of his research into the universality of facial expressions—including a remarkable clash with Margaret Mead—in his afterword to the 1998 edition of *The Expression of the Emotions in Man and Animals.*
25. " 'Duchenne smile' ": Darwin, 1998, 203.
25. "studies of stroke victims": Damasio, 1995, 140.
26. "sometimes called 'modules' ": There is a whole cottage industry of debate about the correct terminology for these "component parts" of the brain. Some prefer terms like "system" or "circuitry," partly because they convey more clearly that modules are usually distributed throughout the brain, and involve both areas of activation and neuromodulatory chemicals. I've generally adopted "modules" in this text because it is a term used both in neurological accounts of the brain's activity as well as evolutionary psychological accounts. The term itself originates with Jerry Fodor, who suggested that modules possessed the following attributes:

   1. domain specificity
   2. encapsulation
   3. obligatory firing
   4. shallow outputs
   5. speed
   6. inaccessibility to consciousness
   7. a characteristic ontogenetic course
   8. a dedicated neural architecture
   9. a characteristic pattern of breakdown

26. "How many million": Woolf, 37.
27. *"the sum of your modules":* For more on modules, see Gardner, page 55. Antonio Damasio is characteristically erudite on the interconnections among the various modular elements. "We can now say with confidence that there are no single 'centers' for vision, or language, or for that matter, reason or social behavior. There are 'systems' made up of several interconnected brain units; anatomically, but not functionally, those brain units are none other than the old 'centers' of phrenologically inspired theory; and these systems are indeed dedicated to relatively separable operations that constitute the basis of mental functions. It is also true that the separate brain units, by virtue of where they are placed in a system, contribute different components to the system's operation and are thus not interchange-

able. This is most important: What determines the contribution of a given brain unit to the operation of the system to which it belongs is not just the structure of the unit but also its *place* in the system." Damasio, 1995, 16.

27. "directly by taking drugs": Psychedelic drugs help undermine two of the most powerful—and, when you think about it, miraculous—effects of our consciousness: that we are a unified self, and not a host of competing subsystems, and that we are each of us distinct from the world around us, that our selves terminate at the peripheries of our bodies. The two trademark effects of psychedelia—ego fragmentation and ego expansiveness—are both evidence of how hard it is for the brain to pull off the illusion that we are one and not many, how easily disturbed these illusions are with the right chemicals. For two very different but equally fearless investigations into these issues, see John Horgan's *Rational Mysticism*, and Daniel Pinchbeck's *Breaking Open the Head*.

28. "control structures between modules": The clash between different modules can sometimes seem like bad engineering, and in a sense it is. Our brains are a reminder that evolution is full of jerry-rigged solutions and inefficient design. Joe LeDoux describes it this way: ". . . there is an imperfect set of connections between cognitive and emotional systems in the current stage of evolution of the human brain. This state of affairs is part of the price we pay for having newly evolved cognitive capacities that are not yet fully integrated into our brains. Although this is also a problem for other primates, it is particularly acute for humans, since the brain of our species, especially our cortex, was extensively rewired in the process of acquiring natural language functions. . . . Our brain has not evolved to the point where the new systems that make complex thinking possible can easily control the old systems that give rise to our base needs and motives, and emotional reactions. This doesn't mean that we're simply victims of our brains and should just give in to our urges. It means that downward causation is sometimes hard work. Doing the right thing doesn't always flow naturally from knowing what the right thing to do is." LeDoux, 2002, 322–23.

29. "left and right hemisphere": "The distribution of white matter to grey is not even throughout the brain—the right hemisphere has relatively more white matter, while the left has more grey. This microscopic distinction is significant because it means that the axons in the right brain are longer than in the left and this means they connect neurons that are, on average, farther away from one another. Given that neurons that do similar things or process particular types of input tend to be clustered together, this sug-

gests that the right brain is better equipped than the left to draw on several different brain modules at the same time. The long-range neural wiring might explain why that hemisphere is inclined to come up with broad, many-faceted, but rather vague concepts. It might also help the right brain to integrate sensory and emotional stimuli (as is required to apprehend art) and to make the sort of unlikely connections that provide the basis of much humour. 'Lateral thinking' would be helped, too, by the neural arrangement in the right brain—the sideways extension of axons even makes the phrase literal rather than figurative. The left brain, by contrast, is more densely woven. The close-packed, tightly connected neurons are better equipped to do intense, detailed work that depends on close and quick cooperation between similarly dedicated brain cells." Carter, 38.

29. *"a jungle in there"*: "The brain giving rise to the mind is a prototypical complex system, one more akin in its style of construction to a jungle than to a computer. This analogy fails at one point: While plants in jungles are selected for during evolution, the jungle itself is not. But the brain is subjected to two processes of selection, natural selection and somatic selection. The result is a subtle and multilayered affair, full of loops and layers. From genes to proteins, from cells to orderly development, from electrical activity to neurotransmitter release, from sensory sheets to maps, from shape to function and behavior, from social communication back to any and all of these levels, we are confronted with a system of somatic selection that is continually subjected to natural selection." Edelman, 1992, 44.

30. "mindreading and eye-reading": ". . . gaze monitoring is seen in infants from around 9 months of age, and which all children, the world over, show by 14 months or so. In this phenomenon, the infant turns in the same direction that another person is looking at and then shows gaze alternation, checking back and forth a few times to make sure (as it appears) that it and the other person are both looking at the same thing, thus establishing shared visual attention on the same object." Baron-Cohen, 44.

32. "Your brain is wired": Many other animals have dedicated brain systems that regulate social interactions. "Dolphins have a massive new brain area, the paralimbic lobe, that we do not possess. The paralimbic lobe is an outgrowth of the cingulate gyrus, which is known to elaborate social communication and social emotions (such as feelings of separation distress and maternal intent) in all other mammals. Thus, dolphins may have social thoughts and feelings that we can only vaguely imagine." Panksepp, 61. "For their size, vampire bats have very big brains. The rea-

son is that the neocortex—the clever bit at the front of the brain—is disproportionately big compared to the routine bits toward the rear. Vampire bats have by far the largest neocortexes of all bats. It is no accident that they have more complex social relationships than most bats, including, as we have seen, bonds of reciprocity between unrelated neighbours in a group. To play the reciprocity game, they need to recognize each other, remember who repaid a favour and who did not, and bear the debt or the grudge accordingly. Throughout the two cleverest families of land-dwelling mammals, the primates and the carnivores, there is a tight correlation between brain size and social group. The bigger the society in which the individual lives, the bigger its neocortex relative to the rest of the brain. To thrive in a complex society, you need a big brain. To acquire a big brain, you need to live in a complex society. Whichever way the logic goes, the correlation is compelling." Ridley, 1996, 69.

32. *"we're all extroverts":* "One might wonder how the selective pressure could have been very strong during recent human evolution. After all, what usually generates the pressure is a hostile environment—droughts, ice ages, tough predators, scarce prey—and as human evolution has proceeded, the relevance of these things has abated. The invention of tools, of fire, the advent of planning and cooperative hunting—these brought growing control over the environment, growing insulation from the whims of nature. How, then, did ape brains turn into human brains in a few million years? Much of the answer seems to be that the environment of human evolution has been human (or prehuman) beings. The various members of a Stone Age society were each other's rivals in the contest to fill the next generation with genes. What's more, they were each other's tools in that contest. Spreading their genes depended on dealing with their neighbors: sometimes helping them, sometimes ignoring them, sometimes exploiting them, sometimes liking them, sometimes hating them—and having a sense for which people warrant which sort of treatment, and when they warrant it. The evolution of human beings has consisted largely of adaptation to one another." Wright, 1995, 26–27

33. "extreme version of the male brain": Baron-Cohen proposes that there is an opposing, traditionally male-centered trait to the empathy of mindreading: systemizing. He describes it as follows: "There are at least six kinds of systems that the human brain can analyse or construct:

1. Technical systems: a computer, a musical instrument, a hammer, etc.
2. Natural systems: a tide, a weather front, a plant, etc.

3. Abstract systems: mathematics, a computer program, syntax, etc.
4. Social systems: a political election, a legal system, a business, etc.
5. Organisable systems: a taxonomy, a collection, a library, etc.
6. Motoric systems: a sports technique, a performance, a technique for playing a musical instrument, etc.

Systemising is an inductive process. You watch what happens each time, gathering data about an event from repeated sampling, often quantifying differences in some variables within the event and their correlation with variation in outcome. After confirming a reliable pattern of association—generating predictable results—you form a rule about how this aspect of the system works. When an exception occurs, the rule is refined or revised; otherwise, the rule is retained." Baron-Cohen, 2002.

40. "instinctive 'gut feeling' ": Antonio Damasio has studied the ways in which those gut feelings enhance and direct our rational assessment of the world, most notably in *Descartes' Error.* He refers to these emotional cues as "somatic markers"—hints from your emotional subsystems that help you navigate complicated situations without having to process everything consciously: "trust this person," "be on the lookout in this neighborhood."

41. "I asked Baron-Cohen": interview conducted January 2003.
41. "unable to detect fearful expressions": Damasio, 1998, 65.
44. "both were seeing": James, 89–90.
44. "cultural achievements of art": This is one place where I think Steven Pinker and E. O. Wilson have it wrong. Here's Pinker from *The Blank Slate:* "Modernism certainly proceeded as if human nature had changed. All the tricks that artists had used for millennia to please the human palate were cast aside. In painting, realistic depiction gave way to freakish distortions of shape and color and then to abstract grids, shapes, dribbles, splashes, and, in the $200,000 painting featured in the recent comedy *Art,* a blank white canvas. In literature, omniscient narration, structured plots, the orderly introduction of characters, and general readability were replaced by a stream of consciousness, events presented out of order, baffling characters and causal sequences, subjective and disjointed narration, and difficult prose." Pinker, 2002, 449.

Wilson is less polemical than Pinker on this point, and he does include a stirring—and I suspect accurate, though who really knows—account of the origins of the arts.

"The arts filled the gap. Early humans invented them in an attempt to express and control through magic the abundance of the environment, the power of solidarity, and other forces in their lives that mattered most to survival and reproduction. The arts were the means by which these forces could be ritualized and expressed in a new, simulated reality. They drew consistency from their faithfulness to human nature, to the emotion-guided epigenetic rules—the algorithms—of mental development. They achieved that fidelity by selecting the most evocative words, images, and rhythms, conforming to the emotional guides of the epigenetic rules, making the right moves. The arts still perform this primal function, and in much the same ancient way. Their quality is measured by their humanness, by the precision of their adherence to human nature. To an overwhelming degree that is what we mean when we speak of the true and beautiful in the arts." Wilson, 225. Wilson's story makes perfect sense in the context of myth, but as a theory of art it falls short—precisely because art is as much about the grain of individual experience as it is about human universals. Neither Pinker nor Wilson seem willing to accept in the arts what they regularly accept in other domains: that part of cultural achievement is about breaking free from the chains of our biology, pushing the boundaries of what human nature is capable of. The fact that Pinker includes the stream of consciousness as part of modernism's break from human nature seems particularly odd to me, since Joyce's whole project was to capture the interior, lived experience of consciousness, in all its dynamism and strangeness. This of course is consonant with much of literary modernism, which is why this book returns several times to Henry James, Virginia Woolf, and Marcel Proust—all of whom were brilliant "mind openers."

45. "taking hold of experience": Woolf, page 79. "One object, one word, one glance, will act like a stone cast into a puddle, sending out ripples of memories triggering off one another. Gradually, beyond the generic concepts, the verbal label, we acquire a more sophisticated and highly personalized view, and understanding of a more sophisticated and highly personalized view, and understanding of the world that is unique to each of us. As we grow up, we are better able to understand or explain an ongoing situation in the light of previous experience, and objects, people, or actions with only a modest impact on the raw senses will monopolize our attention because of a far more covert, less tangible, and private significance. We start to register not only the loudest and brightest in each waking moment, but the secret lover's eyebrow raised silently for a frac-

tion of an inch over a fraction of a second across a noisy, crowded room."
Greenfield, 54.

45. "*all* the chapters are about memory": One of Peter Kramer's long-decay
ideas follows exactly this logic. "We readily accept the notion of cognitive
and emotional, or at least emotion-laden, memory. But perhaps sensitivity is memory as well—'the memory of the body,' as we might say 'the wisdom of the body.' In this sense, social inhibition and rejection-sensitivity
are both memory. That is, they do not stem from a (cognitive, emotion-
laden, conflicted) memory of trauma; they represent or just are memories
of trauma. According to this way of thinking, much of who Lucy is—her
neural pathways, her social needs—constitutes a biological memory of her
mother's murder, just as Tess's social style is a memory of her precociously
responsible childhood." Kramer, 124.

46. "a process called reconsolidation": "The recent discovery, made by Karim
Nader and Glenn Schafe in my lab, is that protein synthesis in the amygdala seems necessary for a recently activated memory to be kept as a
memory. That is, if you take a memory out of storage you have to make
new proteins (you have to restore, or reconsolidate it) in order for
the memory to remain a memory. One way of thinking about this is that
the brain that does the remembering is not the brain that formed the initial memory. In order for the old memory to make sense in the current
brain, the memory has to be updated. This work has stimulated a lot of
interest from both scientists and lay persons. One man called and asked
whether it might be possible for him to eliminate the memory of his ex-
wife by blocking protein synthesis in his brain while thinking of her. The
practical side of this is that it might be possible some day to have trauma
victims recall their trauma in the presence of some drug or other brain
alteration that reduces the stranglehold of the memory on the person's
psyche. After we proposed this, though, a therapist made a very good
point. What would it mean to a Holocaust survivor, for example, to lose
such memories after having lived for many years and having developed an
identity based in part on them? This is a very important concern, and
touches on the deep ethical issues that scientific discoveries can raise."
LeDoux, 2002, 162.

46. "Proust": The writer Stephen Hall published a wonderful essay in *The
New York Times* several years ago that partly inspired some of the themes
of this book. (By accident, I ended up using the same guide for a personal
fMRI exploration of my brain, Columbia's Joy Hirsch.) Hall has this to
say about the mind-opening possibilities of brain-imaging technology:

"A common thread of both the Freudian and Proustian worldviews is the associative quality of recollection—the odd word or sight that connects to a deeper trauma, the odor that connects to a more extensive memory. Association requires connections, and as I saw, a brain scan of humor, for example, can actually depict a rich skein of associations in a diagram of neural connections. Preposterous as it may seem, I can imagine a day in the distant future when the M.R.I. replaces the couch, when the therapist uses words or odors or pictures to excite and pinpoint circuitry and then the neuroanatomist translates the images into explanations of behavior. Of course, there is always the possibility that after decades of exploration in search of mind, we'll still find ourselves, metaphorically speaking, knee-deep in a swamp of neurotransmitters that may bring us no closer to a biological understanding of 'mind.'" Hall, 1999.

2: THE SUM OF MY FEARS

51. "Edouard Claparede": LeDoux, 1996, 180.
54. "My first grant": interview conducted with Joseph LeDoux, November 2002.
54. "can't be studied scientifically": Damasio describes the indifference to emotion this way: "Twentieth-century science left out the body, moved emotion back into the brain, but relegated it to the lower neural strata associated with ancestors whom no one worshiped. In the end, not only was emotion not rational, even studying it was probably not rational." Damasio, 1998, 39.
54. "surgical subtraction": "Studies by Liz Romanski, Claudia Farb, Neot Doron and me show that the lateral amygdala gets inputs about the stimuli from two sources. It receives a crude but fast representation from a subcortical area (the sensory thalamus) and a slower but more complete representation from cortical sensory areas. . . . The role of these two input systems to the amygdala was elucidated in studies performed in my lab by Liz Romanski. The path from the thalamus through the cortex to the amygdala, the so-called high road, allows complex information about objects and experiences to initiate fear reactions. But the amygdala also can be activated directly from the thalamus. Since this low road bypasses the neocortex, it only provides the amygdala with a crude representation of the external stimulus. But the arrival of crude information can have important consequences. For example, Fabio Bordi and I found that cells

in the amygdala are able to determine the intensity or loudness of a sound through the thalamic pathway. Loudness is a good clue to how close something is and distance is a good clue to how dangerous that thing is. If you treat loud things as dangerous even if you don't know the source of the noise, you are probably going to be better off in the long run. So simply by computing intensity from the thalamus, the amygdala can immediately deduce significant details about a stimulus. Intensity is not the only feature gauged by the low road from the thalamus, but it's an important one." LeDoux, 2002, 122.

58. " 'Hebbian learning' ": "In Hebb's words, 'When an axon of cell A is near enough to excite cell B or repeatedly and consistently takes part in firing it, some growth process or metabolic changes take place in one or both cells such that A's efficiency, as one of the cells firing B, is increased.' Although Hebb originally proposed his fire-and-wire theory to account for the nature of learning and memory, it has been used to explain other aspects of synaptic function, especially the construction of synapses during development. Consider again the establishment of visual cortex connectivity. As we've seen, in primates, the weeks before birth are an important period for visual system development, as waves of spontaneous activity from the two eyes set up patterns of activity that result in the preferential activation of certain cortical cells by one eye or the other. Because cells in the retina of one eye are more likely to fire spontaneously at the same time, and are much less likely to fire at the same time as cells in the other eye, chances are that when a postsynaptic cortical cell is activated by presynaptic inputs from one eye, presynaptic inputs from other cells in the same eye will be arriving more or less simultaneously. According to Hebb's rule, this concurrent activity in presynaptic and postsynaptic cells then leads to a strengthening of the connections from that eye to the postsynaptic cell." LeDoux, 2002, 79–80.

61. "flashbulb memories stored": "According to the interleaved learning hypothesis, then, the memory is initially stored via synaptic changes that take place in the hippocampus. When some aspects of the stimulus situation recur, the hippocampus participates in the reinstatement of the pattern of cortical activation that occurred during the original experience. Each reinstatement changes cortical synapses a little. Because the reinstatements depend on the hippocampus, damage to the hippocampus affects recent memories, but not old ones that have already been consolidated in the cortex. Old memories are the result of accumulations of synaptic changes in the cortex as a result of multiple reinstatements of the

memory. The slow rate of change of the cortex prevents the acquisition of new knowledge from interfering with old cortical memories. Eventually, the cortical representation comes to be self-sufficient. At that time, the memory becomes independent of the hippocampus." LeDoux, 2002, 106.

61. "James McGaugh": interview conducted November 2002.

63. "Beta-blockers": "These results raise the intriguing possibility that beta-blockers such as propranolol could be administered to trauma survivors in order to reduce persisting memories. Beta-blockers might also be given ahead of time to emergency workers before they enter a disaster site, and thus thwart altogether the development of intrusive memories that will plague them later. These are exciting possibilities because intrusive memories can be so crippling for long periods of time. And for emergency or disaster personnel who are repeatedly exposed to potential sources of persistence, preliminary administration of beta-blockers might make a highly stressful occupation more manageable." Schacter, 182.

64. "deliberate overriding of fear responses": LeDoux, 1996, 169.

66. "severe stress may impede": See McEwen and Sapolsky, 1995.

3: YOUR ATTENTION, PLEASE

72. "Neurofeedback dates": See Robbins's *Symphony in the Brain* for an authoritative account of neurofeedback's history.

73. "Attention Builder CEO": interview conducted August 2001.

74. "waves of electrical activity": "At present, five general categories of brain waves are recognized in humans. The slowest rhythm is delta (0.5–3 Hz), which generally tends to reflect that the subject is sleepy. . . . The next is theta (4–7 Hz), which has been related to meditative experiences, unconscious processing, and some negative emotional effects such as frustration. However, as mentioned, theta reflects active information processing in certain brain areas such as the hippocampus (HC). When this rhythm occurs in the HC, an organism is typically exploring and the HC is presumably elaborating thoughts and memories. This rhythm is also characteristic of the HC during rapid eye movement (REM) sleep. . . . The brain's relaxed, or 'idling,' rhythm is alpha (8–12 Hz), which provides an excellent reference measure for detecting changes in brain arousal. In other words, the ongoing electrical energy in the alpha range can be used as a baseline for detecting how various brain areas become aroused during specific cognitive tasks and emotional situations. Beta rhythm (typi-

cally 13–30 Hz) is generally considered an excellent measure of cognitive and emotional activation. Finally, oscillations above beta are usually considered to be in the gamma range (i.e., more than 30 Hz); they are presently thought to reflect some of the highest functions of the human brain, such as perceptual and higher cognitive processes." Panksepp, 87.

79. "Leslie Seiden and Hal Rosenblum": interview conducted September 2001.

83. "number of studies": See Sime, et al, 2001.

86. "John Donoghue": interview conducted September 2001.

87. "alpha states": "There is already one technology that appears to generate at least one aspect of a spiritual experience. This experimental technology is called Brain Generated Music (BGM), pioneered by NeuroSonics, a small company in Baltimore, Maryland, of which I am a director. BGM is a brain-wave biofeedback system capable of evoking an experience called the Relaxation Response, which is associated with deep relaxation. The BGM user attaches three disposable leads to her head. A personal computer then monitors the user's brain waves to determine her unique alpha wavelength. Alpha waves, which are in the range of eight to thirteen cycles per second (cps), are associated with a deep meditative state, as compared to beta waves (in the range of thirteen to twenty-eight cps), which are associated with routine conscious thought. Music is then generated by the computer, according to an algorithm that transforms the user's own brain-wave signal. The BGM algorithm is designed to encourage the generation of alpha waves by producing pleasurable harmonic combinations upon detection of alpha waves, and less pleasant sounds and sound combinations when alpha detection is low. In addition, the fact that the sounds are synchronized to the user's own alpha wavelength to create a resonance with the user's own alpha rhythm also encourages alpha production." Kurzweil, 157.

89. "John Rodenbough": interview conducted February 2003.

90. "the 'user illusion' ":

91. " 'phonological loop' ": "We now know that the kind of rapid transience associated with a damaged phonological loop has significant, even grave consequences. The early clues came from studies of another brain-injured patient with a damaged phonological loop. The patient could learn word pairs in her native language, Italian, as quickly as healthy controls. But in contrast to healthy native Italian speakers, the patient could not learn Italian words paired with unfamiliar Russian words. Subsequent studies showed similar results: patients with damage to the phonological loop were almost totally unable to learn foreign vocabulary. The phonological

loop turns out to be a gateway to acquiring new vocabulary. The loop helps us put together the sounds of novel words. When it is not functioning properly, we cannot hold on to those sounds long enough to have a chance of converting our perceptions into enduring long-term memories." Schacter, 30.

91. "Encoding is the attention subsystem": "The other region whose activity predicted subsequent memory was located farther forward, in the lower left part of the vast territory known as the frontal lobes. This finding was not entirely unexpected, because previous neuroimaging studies indicated that the lower left part of the frontal lobes works especially hard when people elaborate on incoming information by associating it with what they already know. Cognitive psychologists had known for years that transience is influenced by what happens as people register or encode incoming information: more elaboration during encoding generally produces less transient memories. For instance, suppose I show you a list of words to remember, including lion, CAR, table, and TREE. For half of the words, I ask you to judge whether they refer to living or nonliving things; for the other half, I ask you to judge whether they are in uppercase or lowercase letters. All other factors being equal, you will later remember many more of the words for which you had made living/nonliving judgements. Thinking about whether the word refers to a living or nonliving thing allows you to elaborate on the word in terms of what you already know about it; making the uppercase/lowercase judgement does little to link the word with what you already know. Other experiments have shown that subsequent memory improves when people generate sentences or stories that tie together to-be-learned information with familiar facts and associations." Schacter, 25.

91. "seven distinct items": "It has been known for centuries that we can only keep a few things active in our mind (in working memory) at once. George Miller, one of the pioneers in cognitive psychology, figured out, through psychological experiments, that the number is about seven pieces of information. Some people can hang on to eight or nine, whereas others manage only five, but, on average, temporary storage can hold about seven items. (It's probably no coincidence that telephone numbers within an area code were designed to have seven digits.) But, as Miller noted, we can effectively expand that capacity by chunking or grouping information—it's about as easy to remember seven letters as seven words or ideas. No doubt one of the reasons human cognition is so powerful is because we have language in our brains, which exponentially increases

the ability to categorize information, to chunk. A whole culture, for instance, can be implied by a name." LeDoux, 2002, 177.

99. "my visual encoding": This is a trait that I seem to share with Aldous Huxley. "I am and, for as long as I can remember, I have always been a poor visualizer. Words, even the pregnant words of poets, do not evoke pictures in my mind. No hypnagogic visions greet me on the verge of sleep. When I recall something, the memory does not present itself to me as a vividly seen event or object. By an effort of the will, I can evoke a not very vivid image of what happened yesterday afternoon, of how the Lungarno used to look before the bridges were destroyed, of the Bayswater Road when the only buses were green and tiny and drawn by aged horses at three and a half miles an hour. But such images have little substance and absolutely no autonomous life of their own. They stand to real, perceived objects in the same relation as Homer's ghosts stood to the men of flesh and blood, who came to visit them in the shades." Huxley, 15.

100. "effective brain": Another area that is key to "tool organization" is the association cortex, which draws on inputs from the higher-order somatic, visual, and auditory areas. This region plays a central role in creating an integrated assessment of one's immediate environment, drawing connections between the different senses.

101. "Susan and Sigfried Othmer": interview conducted December 2002.

4: SURVIVAL OF THE TICKLISH

108. "oxytocin": "Oxytocin is probably best known for what it contributes to birth, prompting labor itself and milk production. The sensations that accompany the release of oxytocin hold special interest. Right after birth, an intense calm sets in for most mothers. You've just completed one of the most vigorous and painful experiences of your life, which lasted for perhaps ten or fifteen hours, and it really is nice to have it over. But the calm is more than what comes from relief at the end of a painful experience. It has an otherworldly quality. When you look at paintings of the Madonna, you get the sense that some artists have crept into the new mother's soul and sensed what those feelings are really like. Certainly love for the newborn is part of it, but the intensity is greater and more visceral than love connotes. This is the beginning of bonding." Taylor, 25.

108. "Shelley Taylor": interview conducted December 2002.

110. "'The Tending Instinct'": Taylor also calls this the affiliative circuitry. "Does our legacy of tending to one another's needs truly merit consideration as an instinct? Can we argue with any confidence that there is a biologically driven program that underlies the many relationships in which we nurture one another? As we explore the nature of our social ties, first in the mother-infant relationship, then in relations within a social group, and between women and men, some of the same hormones will appear repeatedly—oxytocin, vasopressin, endogenous opioid peptides, growth hormone—among other suggestive commonalities. These hormones appear to be implicated in social behaviors of many kinds and are part of what scientists have called the affiliative neurocircuitry, an intricate pattern of co-occurring and interacting pathways that influence many aspects of social behavior, ranging from whether people are receptive to social relationships at all to how strong their relationships will be." Taylor, 12.

111. "testosterone-heavy male bodies": Oxytocin has something of an analog in the male brain—a hormone called vasopressin. "Both men and women release vasopressin in response to stress, but unlike oxytocin, which is subdued by male hormones, the effects of vasopressin may be enhanced by them, making vasopressin a potential influence on men's stress responses. If oxytocin is associated with calm, nurturant, affiliative behavior, what does vasopressin do? Again, because much of this action goes on in the brain, the source of our knowledge about vasopressin is from animal studies. One animal in particular, the prairie vole, has provided a lot of knowledge. Why the prairie vole and not rats, rhesus monkeys, and sheep, which provided help in understanding the effects of oxytocin in women? Unlike most male mammals, the prairie vole is a monogamous little creature who picks a mate and stays with her for life. He guards and protects her and generally keeps things safe. Since humans are fairly monogamous too, the prairie vole provides a potential animal model for understanding what men do in response to stress. The little bit of research that has been done suggests that vasopressin may well be implicated in male responses to stress. When stress occurs, levels of vasopressin go up, and the male prairie vole becomes a protective sentinel, guarding and patrolling his territory, keeping his female partner and the young from harm." Taylor, 31.

112. "brains of prairie voles": LeDoux, 231.

112. "brain chemistry of humans": Panksepp has some characteristically thought-provoking ideas about the impact of these chemicals on male-

female relations. ". . . the female brain contains more oxytocin neurons than the male brain, and the genetic manufacturer of oxytocin is under the control of the ovarian hormone estrogen. The role of this neuropeptide in sexuality is not as lopsided as that of vasopressin in the male brain. Administration of oxytocin directly into the brain can increase both male and female sexuality, but seemingly in different ways. In males, oxytocin promotes erectile capacity, and it is released into the circulation in large amounts at orgasm. . . . Unfortunately, no comparable data appear to be available for females. In any event, at present, brain oxytocin release is a key candidate for being a promoter of orgasmic pleasure and hence one of the mediators of behavioral inhibition commonly seen in males following copulation. There is a certain beauty in the fact that oxytocin, a predominantly female neuromodulator, is an especially important player in the terminal orgasmic components of male sexual behavior. In that role it may allow the sexes to better understand each other." Panksepp, 241.

112. "human oxytocin receptors": "It is noteworthy that no neurotransmitter or neuromodulator has been discovered in humans that is qualitatively different from those found in other mammals. In fact, all mammals share remarkably similar anatomical distributions of most neurochemical systems within their brains. However, there are also distinctions in systems between different animals and species, which help explain their personality differences. One of the most dramatic contrasts observed so far is within brain oxytocin systems . . . of various species." Panksepp, 100.

115. "Reptile brains do not produce": "The emotional tendency to provide special care to the young, so impressive in mammals, is seen only in rudimentary forms in reptiles. Still, a primitive tendency to provide maternal care probably evolved before the divergence of mammalian and avian stock from their common ancestor. This is suggested by the strong parental urges of most avian species and by recent paleontological evidence suggesting that some dinosaurs may also have exhibited maternal tendencies. However, maternal devotion, through the evolution of CARE systems, has vastly expanded within the mammalian brain, while remaining rooted in the sociosexual processes that had evolved earlier." Panksepp, 223.

115. "The evolutionary biologist Donald Symons": Pinker expands on Symons's argument with this thought experiment: "One can even imagine a species in which every couple was marooned on an island for life and their offspring dispersed at maturity, never to return. Since the

genetic interests of the two mates are identical, one might at first think that evolution would endow them with a blissful perfection of sexual, romantic, and companionate love. But, Symons argues, nothing of the sort would happen. The relation between the mates would evolve to be like the relation among the cells of a single body, whose genetic interests are also identical. Heart cells and lung cells don't have to fall in love to get along in perfect harmony. Likewise, the couples in this species would have sex only for the purpose of procreation (why waste energy?), and sex would bring no more pleasure than the rest of reproductive physiology such as the release of hormones or the formation of the gametes. There would be no falling in love, because there would be no alternative mates to choose among, and falling in love would be a huge waste. You would literally love your mate as yourself, but that's the point: you don't really love yourself, except metaphorically; you are yourself. The two of you would be, as far as evolution is concerned, one flesh, and your relationship would be governed by mindless physiology. You might feel pain if you observe your mate cut herself, but all feelings we have about our mates that make a relationship so wonderful when it is working well (and so painful when it is not) would never evolve. Even if a species had them when they took up this way of life, they would be selected out as surely as the eyes of a cave-dwelling fish were selected out, because they would be all cost and no benefit." Pinker, 2002, 293–94.

119. "Robert Provine": interview conducted January 2003.

119. "46 percent more likely": "While tabulating the data, I found that speakers laughed more than their audiences. Nothing in the audience-oriented literature about laughter or humor suggested such a result. When I totaled speaker (S) and audience (A) laughter across all four possible gender combinations, speakers were found to laugh 46 percent more than their audiences. The effect was even more striking when gender was considered. The speaker/audience difference was greatest when females were conversing with males (S_fA_m), a condition in which females produced 126 percent more laughter than their male audiences."

Dyad	Episodes	Speaker	Audience
S_mA_m	275	75.6%	60.0%
S_fA_f	502	86.0%	49.8%
S_mA_f	238	66.0%	71.0%
S_fA_m	185	88.1%	38.9%
overall	1,200	79.8%	54.7%

Provine, 28.

119. "only around 15 percent": Provine, 47–48.
123. "their laughter is contagious": Nowhere is the power of contagious laughter more conspicuously exploited than on television laugh tracks, which
first appeared in September of 1950, accompanying *The Hank McCune
Show*—a comedy, Provine writes, "about 'a likeable blunderer, a devilish
fellow who tries to cut corners only to find himself the sucker.' *Variety* (13
September 1950) was alert to the one innovation in this 'fairly amusing'
show aired on NBC—'there are chuckles and yucks dubbed in.' Sensing
something of interest, the reviewer mused, 'Whether this induces a jovial
mood in home viewers is still to be determined, but the practice may have
unlimited possibilities if it's spread to include canned peals of hilarity,
thunderous ovations and gasps of sympathy.' The outcome of this experiment is all too familiar to contemporary television viewers." Provine,
137.
124. "Roger Fouts": interview conducted January 2003.
124. "chimpanzee laughter": "The most notable acoustic similarity between
human and chimpanzee laughter is its rhythmic structure. Whether a
chimp is 'pant-pant-panting,' or a person is saying 'ha-ha-ha,' the sonic
bursts occur at regular intervals, a property apparent in the waveforms of
both vocalizations. Chimps, however, have a laugh rhythm about twice as
fast as that of humans. (The chimp sounds were separated—onset to
onset—by about 120 milliseconds, versus about 210 milliseconds for
humans.) This is because the chimpanzees vocalize during both inhala-
tion and exhalation. If only the more strongly voiced exhalation is considered, the chimpanzee laugh rate is halved and approximates that of
humans." Provine, 79–80.
124. "Parents will testify that ticklefests": Tickling has a surprisingly rich history of philosophical investigation. "Tickle is a strange behavior, but we
need not search for an exotic neural mechanism to explain it. A well-
known neural process accounts for many of tickle's perplexing qualities.
The central clue about the nature of the tickle mechanism is this: We can't
tickle ourselves. Over 2,000 years ago Aristotle showed acute intuition
about this phenomenon: 'Is it because one also feels tickling by another
person less if one knows beforehand that it is going to take place, and
more if one does not foresee it? A man will therefore feel tickling least
when he is causing it and knows that he is doing so.'" Provine, 116.
125. "Jaak Panksepp": interview conducted December 2002.
126. "drugs that block the effects": Panksepp, 256.
126. "immune system": Provine, 197.

128. "a small retreat": Thanks to Clay Shirky for putting together such a fascinating seminar, and for his continued brilliance on the question of software's social possibilities.

130. "cultural critic Harvey Blume": "Autism and the Internet,"
http://web.mit.edu/m-i-t/articles/blume.html.

130. "Sue Carter": interview conducted December 2002.

132. " 'feel crummy' hormone": The idea that endogenous drugs could have
such contradictory effects sometimes seems illogical, but again it's
helpful to think here of recreational drugs. Alcohol in low doses often
makes people more socially confident and outgoing, while in high doses
it functions as a depressant—making people hostile to others, or indifferent to them. The drug in both cases is the same—it's the dose that
matters.

132. "body's natural opiates": "It has been postulated that there are parallels
between social attachment and narcotic addiction and that similar neural
circuitry and neurochemistry may underlie both phenomena. . . . In fact,
there is significant evidence for a role of endogenous opioids in modulating affiliative behaviors. For example, b-endorphin is released in monkeys
during social grooming, while blockade of opioid receptors results in
increased motivation to be groomed. . . . Opiates also modulate infant-
mother attachments and separation distress calls in rat pups. . . . Our
research on the neurobiological mechanisms regulating affiliation and
pair bonding in voles supports the idea that brain reward circuitry plays a
key role in regulating social attachments." Insel, et al., 2001.

132. " 'addicted to love' ": "EOPs [endogenous opioid peptide] may even exert
direct effects on stress hormones, reducing their strength. When animals
are socially isolated, their levels of EOPs decline. When they rejoin their
companions, EOP levels return to normal, accompanied by an emotional
state that can only be described as euphoria. Neuroscientist Jaak Panksepp
has suggested that EOPs may be the key to a mild form of social addiction, whereby the release of opioids in response to companionship sustains
the need for that companionship. So far, only animal evidence for this idea
has accumulated, so it is not yet known whether EOPs underlie the
euphoria that human companionship can produce or the need to seek it
out in the face of loneliness or isolation." Taylor, 83.

133. "the same altar": "The first neurochemical system that was found to exert
a powerful inhibitory effect on separation distress was the brain opioid
system. This provided a powerful new way to understand social attachments. There are strong similarities between the dynamics of opiate

addiction and social dependence . . . and it is now clear that positive social interactions derive part of their pleasure from the release of opioids in the brain. For instance, the opioid systems of young animals are quite active in the midst of rough-and-tumble play, and when older animals share friendly time grooming each other, their brain opioid systems are activated. . . . Finally, sexual gratification is due, at least in part, to opioid release within the brain. From all this, it is tempting to hypothesize that one reason certain people become addicted to external opiates (i.e., alkaloids, such as morphine and heroin, that can bind to opiate receptors) is because they are able to artificially induce feelings of gratification similar to that normally achieved by the socially induced release of endogenous opioids such as endorphins and enkephalins. In doing this, individuals are able to pharmacologically induce the positive feeling of connectedness that others derive from social interactions." Panksepp, 255.

134. "children in their first days of life": "Within the first few hours, he turns his head in the direction of his mother's voice. A social bond that was formed during pregnancy is already in place, and on it, other bonds will be built. After a few hours, the newborn can mimic an adult's expression, and, shortly thereafter, he or she can reciprocate another person's emotions. To demonstrate this amazing capacity, scientists show newborns close-ups of faces that are smiling, frowning, or expressing surprise. They film the babies looking at these pictures and then show the films to observers whose job it is to guess what picture the baby was looking at. Observers are often able to tell, because the baby spontaneously mimics the face in his or her own expression. Infants use this remarkably sophisticated innate system of emotional communication to signal their needs to their caretakers and evoke their tending, and these interactions fuel the exuberant brain growth of infancy." Taylor, 40.

134. "the same chemicals": A fascinating avenue for exploration opened up by this neurochemical approach is the one explored so vividly by Freud: how much overlap is there between filial and sexual love? Panksepp writes: "Filial love—the love between parent and child—seems outwardly quite distinct from sexual desire, but as Freud suspected, they may share important features. As we will see, findings from modern psychobiology can now be used to bolster this view; key molecules such as oxytocin are involved in both, albeit by actions in different parts of the brain. Although our cultural evolution has sought to bind our desire for sex and our need for social bonding together in an inextricable whole called the

institution of marriage, there is no guarantee in the recesses of the brain that such cultural unions will succeed." Panksepp, 226.

But there is a distinction here—the two systems aren't necessarily identical. "One plausible way of thinking is that nurturant love emerges from brain systems that promote parental attachments, while erotic love may emerge from brain systems that generate sexual seeking. If so, the first might be more opiate- and oxytocin-based, while the latter is more dopamine- and vasopressin-based." Panksepp, 285.

134. "essential part of the feeling": Of course, what goes for infants also goes for many of our fellow mammals. It can be daunting to think that the core ingredients of love's neurochemistry are shared between humans and prairie voles. Because love is the source of so many of humanity's highest creative achievements, we like to think that the feeling itself is just as unique. But the commonalities of the brain chemistry—and the commonalities of behavior—suggest that at least some part of love's intoxication is experienced by our fellow mammals.

5: THE HORMONES TALKING

135. "handful of researchers": For an energetic first-person account of this discovery, see Pert, *The Molecules of Emotion*.

136. " 'natural high' ": ". . . we all like our endorphins, and we all do things to get them, ranging from jogging to sex. And when we do those things, our endorphin levels are abnormally high. No doubt rapists feel good at some point during or after their crimes; no doubt that pleasure has a biochemical basis; and no doubt this basis will come to light. If defense lawyers get their way and we persist in removing biochemically mediated actions from the realm of free will, then within decades the realm will be infinitesimal. As, indeed, it should be—on strictly intellectual grounds, at least. There are at least two ways to respond to the growing body of evidence that biochemistry governs all. One is to use the data, perversely, as proof of volition. The argument runs as follows: Of course all these criminals have free will, regardless of the state of their endorphins, blood-sugar levels, and everything else. Because if biochemistry negated free will, then none of us would have free will! And we know that's not the case. Right? (Pause.) Right?" Wright, 1995, 352.

136. "endogenous opioids were not alone": The major recreational drugs can be mapped onto the following neurotransmitters: Ecstasy floods the

brain with excess serotonin. Cocaine increases the availability of dopamine, noradrenaline, and serotonin. Hallucinogens like LSD achieve some of their effects by imitating the serotonin molecule. Amphetamines release dopamine and noradrenaline. Nicotine mimics dopamine molecules, as well as activating the nicotinic receptors. Alcohol and other tranquilizers have a more generalized effect, decreasing the activity of GABA in the brain. Opiates, as their name suggests, pass for the brain's naturally occurring opioids. See Carter, 68.

142. "Learning to recognize": "A viewpoint that is gaining currency in psychiatry, under the rubric 'the functional theory of psychopathology,' is that mental states are best understood first through the consideration of particular mental functions, such as mood, cognition, and perception—and that multifaceted entities such as mental illness or personality should be considered secondary. An assumption of the theory is that variation in functions will turn out to arise from a particular state of one or another neurotransmitter." Kramer, 183.

142. "rejection sensitivity": The term "rejection sensitivity" originates with Donald Klein's research, involving some of the antidepressant drugs developed before Prozac. Here's Kramer describing the condition. "We all react to disappointments, even minor ones. A date stands us up. A colleague makes a cutting remark. . . . Always there is a visceral response: the sinking in the stomach, a feeling of weakness, confusion of thought, a momentary sense of sadness and world-weariness. It will pass, we know, this leaden dullness, but for the moment we are deeply affected. For some this pain is worse than for others—lasts longer, paralyzes more thoroughly. They are not depressed, but they are vulnerable. 'Sensitive' is what we call such people, as in: 'Oh, don't be so sensitive,' or 'She's just overly sensitive.'. . . For the most part, psychiatry has ignored sensitivity as unremarkable—not a category of analysis. In the standard diagnostic manual, there is no category labeled 'sensitive.' But the standard manual is a mere matter of consensus. There are many unofficial ways of mapping human variation, charts highlighting colorful byways that, though they have never made it into the conventional guidebook, promise rewarding vistas. One such conceptual route, a diagnosis that under various names has intrigued biological researchers for decades, may be, with the help of Prozac, on its way to becoming a major thoroughfare. The idea underlying this diagnosis is that certain people are physiologically wired to be deeply sensitive to rejection." Kramer, 70–71.

143. "many forms of brain activity:" "There are good reasons to believe that

this system mediates a relatively homogenous central state function. All motivated and active emotional behaviors including feeding, drinking, sex, aggression, play, and practically every other activity (except sleep), appear to be reduced as serotonergic activity increases. However, the conclusion that serotonin mediates behavioral inhibition is tempered by the discovery of a vast diversity of distinct serotonin receptors. At this writing, the number of 5-HT receptors stands at 15. When serotonin acts on certain receptors, emotional behaviors such as anxiety (as measured by behavioral inhibition) increase, but when other receptors are involved, emotionality is reduced. Why such complexity exists on the postsynaptic side, with comparative simplicity at the presynaptic side, remains perplexing. In other words, these systems release a single transmitter, 5-HT, rather globally in the brain, but this substance can operate on a vast number of receptors with apparently very different functional properties. One possible way to make sense of this is that at various synaptic fields, serotonin release is also controlled by local presynaptic mechanisms (i.e., via axo-axonic synapses). Through such local controls, it is possible to have regionally restricted release of 5-HT onto only a subset of serotonin receptors." Panksepp, 111.

144. " 'magic bullet' chemicals": Part of the complexity here comes from the fact that the various neuromodulators have effects on one another, as we saw in the example of oxytocin and the endorphins. There is also the question of *where* the activity is happening: serotonin release in one part of the brain has a very different effect from serotonin release in another. As Damasio writes: "When it comes to explaining behavior and mind, it is not enough to mention neurochemistry. We must know whereabouts the chemistry is, in the system presumed to cause a given behavior. Without knowing the cortical regions or nuclei where the chemical acts within the system, we have no chance of ever understanding how it modifies the system's performance (and keep in mind that such understanding is only the first step, prior to the eventual elucidation of how more fine-grained circuits operate). Moreover, the neural explanation only begins to be useful when it addresses the *results* of the operation of a given system on yet another system. The important finding described above should not be demeaned by superficial statements to the effect that serotonin alone 'causes' adaptive social behavior and its lack 'causes' aggression. The presence or absence of serotonin in specific brain systems having specific serotonin receptors does change their operation; and such change, in turn, modifies the operation of yet other systems, the result of which will

ultimately be expressed in behavioral and cognitive terms." Damasio, 1995, 77.

144. "PET scans on people's brains": Damasio, 1998, 60–62.

145. " 'mood congruity' ": "According to the mood congruity hypothesis, memories are more easily retrieved when the emotional state at the time of memory formation matches the state at the time of retrieval. For example, we are more likely to remember sad than happy events when depressed. Perhaps amygdala activation during retrieval facilitates remembrance by re-creating, at least in part, the emotional state (the state of the brain resulting from amygdala activation, and all its consequences, as discussed above) that occurred during the original experience—the more similar the pattern of activation is during learning and retrieval, the more efficient retrieval is likely to be." LeDoux, 2002, 222.

146. "revolutionary new treatment": Damasio, 2003, 56.

148. "psychologist Kevin Ochsner": Schacter, 164.

151. "production of dopamine": "Dopamine cell bodies are located in the brain stem, in a region called the ventral tegmental area. The axons of these cells then branch extensively and reach many areas of the forebrain, including the prefrontal cortex, where their terminals release dopamine. In primates, the dopamine terminals are fairly evenly distributed throughout the layers, allowing dopamine to bind to receptors and then modulate excitatory and inhibitory transmission in both the input and output layers. Although there are many subclasses of dopamine receptors, the D1 family (which includes D1 and D5 receptors) has been most clearly implicated in working memory. These receptors are located on the spines and shafts of dendrites of excitatory cells and seem to reduce the transfer of excitation from the dendrites to the cell bodies, allowing only especially strong excitatory inputs to get through to the cell bodies and elicit excitation. Dopamine release in the prefrontal cortex also seems to facilitate GABA inhibition, possibly by way of presynaptic facilitation of transmitter release, leading to a further reduction of excitation through prefrontal circuits. Some of these effects appear to involve the triggering of protein kinase A in cells containing dopamine receptors. Integrating these findings, Amy Arnsten has proposed that dopamine participates in working memory by biasing cells to mainly respond to strong inputs and thereby focusing attention on active current goals and away from distracting stimuli." LeDoux, 2002, 189.

151. "the brain's 'pleasure' drugs": The dopamine system was being triggered in the now legendary story of the 1960s experiment that gave a rat the option

of stimulating a part of his brain by pushing a lever. As everyone knows, the rat happily gave up food and drink to push the lever all day, which led researchers to assume that the dopamine system was all about pleasure. But over time, brain scientists and psychologists began to wonder why people who had excess dopamine—schizophrenics, for instance—didn't seem particularly ecstatic. Slowly the theory of dopamine as reward accountant and motivator—rather than pleasure drug—began to emerge out of those reconsiderations. See Sejnowski and Quartz's excellent *Liars, Lovers, and Heroes* for more on this.

152. "pleasure accountant": "Just as brain stimulation reward was initially thought to be due to activation of pleasure centers, dopamine was believed to be the chemical of pleasure. However, as we've seen, the hedonistic (subjective pleasure) view of brain stimulation reward is incorrect, and the hedonistic interpretation of dopamine's role in reward is incorrect as well. For example, blockade of dopamine interferes with instrumental responses motivated by a sweet reward but does not alter the actual consumption of the tasty stuff when it is obtained—the animals still 'like' the reward when they consume it, but they are no longer motivated to work for it. Dopamine is thus more involved in anticipatory behaviors (looking for food or drink or a sexual partner) than in consummatory responses (eating, drinking, having sex). But being hungry or thirsty is unpleasant. Pleasure, to the extent it is experienced . . . would not come during the anticipatory state but instead during consumption. Since dopamine is involved only in the anticipatory phase, and not in the consummatory phase, its effects (at least in the case of primary need states) cannot be explained in terms of pleasure." LeDoux, 2002, 246.

152. " 'seeking' circuitry": "This system makes animals intensely interested in exploring their world and leads them to become excited when they are about to get what they desire. It eventually allows animals to find and eagerly anticipate the things they need for survival, including, of course, food, water, warmth, and their ultimate evolutionary survival need, sex. In other words, when fully aroused, it helps fill the mind with interest and motivates organisms to move their bodies effortlessly in search of the things they need, crave, and desire. In humans, this may be one of the main brain systems that generate and sustain curiosity, even for intellectual pursuits. This system is obviously quite efficient at facilitating learning, especially mastering information about where material resources are situated and the best way to obtain them. It also helps assure that our

bodies will work in smoothly patterned and effective ways in such quests." Panksepp, 52.

156. "British civil service": Ridley, 1999, 155.

157. "wider world": Shelley Taylor has some provocative thoughts on the way social inequality affects stress levels. ". . . social class hierarchies unravel the social fabric. Every relationship is put under strain and suffers as a result, from ties between parents and children to relations between coworkers and friends. When people simply do not have what they need to get by—and, at least as important, observe that others do—then social institutions and relationships become yet another source of strain, rather than the supportive resources they would otherwise be. These problems worsen as the gap between rich and poor widens, and people pay a high price to live in a society that tolerates these gaps. Sociologist Richard Wilkinson has shown that beyond a certain basic income, your health is influenced more by the gap between rich and poor than by your absolute income. One way to see this is by comparing the death rates of countries that have small gaps between the rich and poor with those that have large gaps. For example, Cuba and Iraq are both poor nations with an equivalent gross domestic product (per capita) of $3,100, but the gaps between rich and poor are much smaller in Cuba than in Iraq. Accordingly, people in Cuba live a full 17.2 years longer than people in Iraq. The United States is a much wealthier nation than Costa Rica, yet in Costa Rica, where income gaps are small, life expectancy is higher." Taylor, 184.

157. "long drama of human history": Economists have begun to explore the world of "rational choice" through the lens of brain science, and the historians can't be too far behind. One potential example: the role of dopamine and the brain's novelty system in the collective trauma that was 9/11. Most dramatic historical events arrive in one of two ways, both of which are unlikely to trigger a dopamine-modulated response. They arrive slowly, with a long lead time: the Vietnam War or the Watergate affair. The news builds over months or years, and in some sense it's the duration of the story that marks its importance. Other dramatic events have already happened by the time you hear of them: the *Titanic* has sunk, the *Challenger* has exploded. The news itself is shocking, of course, but what follows is all aftershock.

But think of the sequence on 9/11. A plane has crashed into the World Trade Center. Surprising news, to say the least. But then the shocking twist: a second plane hits. Then there's news of other planes that have stopped responding to air traffic control. Then a dark cloud begins to billow out above the Pentagon. Then a fourth plane is missing somewhere over Pennsylvania. Then the south tower falls. Then the north.

The events of 9/11 came out of nowhere, but perhaps the most important thing was that they kept coming. Instead of turning on *The Today Show* to discover some startling event that had already run its course, we experienced an entire sequence of startling events, in real time. That sequence created, in a true sense, a global dopamine rush. If "the events of 9/11" had been instead a single event with more casualties—a solo plane topples a skyscraper directly, say—it's likely that the act would have had the same geopolitical ramifications. But I suspect it wouldn't have created the same psychic scars, the same flashbacks. The structure of the attack was perfectly designed to create the most searing memory possible in the human brain: mixing abject fear with repeated novelty.

It's interesting to note that the other epic I'll-never-forget-where-I-was trauma of the postwar era—the Kennedy assassination—followed a similar pattern: the President has been shot; then the President is dead; then his accused assassin is arrested; then the assassin is killed as well, on live TV. (On a more lowbrow note, the first few days of the O.J. saga unfolded in a comparable way.) Think of the scars and the fascination left by these events as the trace of dopamine written into our public history.

6: SCAN THYSELF

158. "suffered a stroke": Orenstein, 105.
159. "musical information": "In most tests with normal individuals, musical abilities turn out to be lateralized to the right hemisphere. For example, in tests of dichotic listening, individuals prove better able to process words and consonants presented to the right ear (left hemisphere), while more successful at processing musical tones (and often other environmental noises as well) when these have been presented to the right hemisphere. But there is a complicating factor. When these, or more challenging tasks, are posed to individuals with musical training, there are increasing left hemisphere, and decreasing right hemisphere, effects. Specifically, the more musical training the individual has, the more likely he will draw at least partially upon the left hemisphere mechanisms in solving a task that the novice tackles primarily through the use of right hemisphere mechanisms." Gardner, 119.

161. "appreciation of music": There have been a number of fascinating studies into the intricacies of birdsong. "There are nine thousand bird species, and song learning arises in only three of the twenty-seven major avian groups—parrots, hummingbirds, and oscines. Within this elite class of vocal learners, there are species differences in style of singing and in the details of the learning process. For some, such as the white-crowned sparrow and zebra finch, only one song dialect is learned during early development and then precisely reproduced during each mating season. Others, such as canaries and warblers, create new song variants each season in much the same way that Wagner created thematic variations or leitmotivs during such operatic masterpieces as *The Ring*. In both single- and multiple-dialect species, different populations maintain long-lasting song traditions, themes passed down from generation to generation. The final class of song learners consists of the great mimics, species such as the mockingbirds, lyrebirds, and starlings. These species build an impressive repertoire of sounds, including songs from the local fauna as well as sounds from some of the inanimate objects in the vicinity. In the London area, a chaffinch learned to reproduce the ring of the British telephone company, and then appeared to use it as a prank to cause the master of the house to rush inside." Hauser, 118–19.

161. "musical chills": "Our overriding assumption is that ultimately our love of music reflects the ancestral ability of our mammalian brain to transmit and receive basic emotional sounds that can arouse affective feelings which are implicit indicators of evolutionary fitness. In other words, music may be based on the existence of the intrinsic emotional sounds we make (the animalian prosodic elements of our utterances), and the rhythmic movements of our instinctual/emotional motor apparatus, that were evolutionarily designed to index whether certain states of being were likely to promote or hinder our well-being. However, upon such fundamental emotional capacities, artists can construct magnificent cognitive structures of sound musical cultures that obviously go far beyond any simple affective or evolutionary concerns." Bernatzky and Panksepp, 2002.

161. "pleasure of parenting": "Jaak Panksepp thinks the emotion-tugging effect of certain types of music lies in its similarity to vocal (but not verbal) signals that carry emotional messages between animals. The tension-building sequence with delayed resolution that typically brings about the chilly spine feeling, for example, has features in common with the sounds made by infants—both human and animal—when they are parted from

their mothers. In animals these cries have been found to trigger a drop in oxytocin—the brain chemical most closely associated with parental bonding—and they also bring about a drop in the mother's body temperature. When the mother is reunited with her baby, the child responds by 'resolving' the cry—a vocal performance not dissimilar to closing a phrase of music with a satisfying final note. At the same time the mother's oxytocin level goes up, and her body becomes warmer. Women have been found to feel the tingle more keenly than men, which fits in neatly with this theory." Carter, 148.

163. "Joy Hirsch": interviews conducted in April and May of 2003.

179. "medial frontal gyrus": Interestingly, when the writer Stephen Hall conducted a similar experiment with Hirsch, there was comparable activity in the medial frontal gyrus, though his overall distribution of activity was different from mine.

CONCLUSION: MIND WIDE OPEN

186. "Eric Kandel": Kandel, 1999.

187. "obtain fresh pleasure": Freud, 1961, 8.

192. "belonging to the past": Freud, 1961, 19.

192. "conscious sense of self": "The unconscious, in the narrow meaning in which the word has been etched in our culture, is only a part of the vast amount of processes and contents that remain nonconscious, not known in core or extended consciousness. In fact, the list of the 'not-known' is astounding. Consider what it includes:

1. all the fully formed images to which we do not attend;
2. all the neural patterns that never become images;
3. all the dispositions that were acquired through experience, lie dormant, and may never become an explicit neural pattern;
4. all the quiet remodeling of such dispositions and all their quiet renetworking that may never become explicitly known; and
5. all the hidden wisdom and know-how that nature embodied in innate, homeostatic dispositions." Damasio, 1998, 228.

194. "With incest": "The issue can be drawn more sharply by distinguishing the two principal hypotheses that compete for the explanation of human

incest avoidance. The first is Westermarck's, which I will now summarize in updated language: People avoid incest because of a hereditary epigenetic rule of human nature that they have translated into taboos. The opposing hypothesis is that of Sigmund Freud. There is no Westermarck effect, the great theoretician insisted when he learned of it. Just the opposite: Heterosexual lust among members of the same family is primal and compelling, and not forestalled by any instinctive inhibition. In order to prevent such incest, and the consequent disastrous ripping apart of family bonds, societies invent taboos." Wilson, 178.

Evidence seems overwhelming that the prohibition against incest is a "human universal" and is thus somewhere grounded in our biology rather than being imposed on us by culture. Consider this amazing study of "minor marriages" in which "unrelated infant girls are adopted by families, raised with the biological sons in an ordinary brother-sister relationship, and later married to the sons. The motivation for the practice appears to be to insure partners for sons when an unbalanced sex ratio and economic prosperity combine to create a highly competitive marriage market. Across four decades, from 1957–1995, Wolf studied the histories of 14,200 Taiwanese women contracted for minor marriage during the late nineteenth and early twentieth centuries. The statistics were supplemented by personal interviews with many of these "little daughters-in-law," or *sim-pua,* as they are known in the Hokkien language, as well as with their friends and relatives. What Wolf had hit upon was a controlled—if unintended—experiment in the psychological origins of a major piece of human social behavior. The *sim-pua* and their husbands were not biologically related, thus taking away all of the conceivable factors due to close genetic similarity. Yet they were raised in a proximity as intimate as that experienced by brothers and sisters in Taiwanese households. The results unequivocally favor the Westermarck hypothesis. When the future wife was adopted before thirty months of age, she usually resisted later marriage with her de facto brother. The parents often had to coerce the couple to consummate the marriage, in some cases by threat of physical punishment. The marriages ended in divorce three times more often than "major marriages" in the same communities. They produced nearly 40 percent fewer children, and a third of the women were reported to have committed adultery, as opposed to about 10 percent wives in major marriages. In a meticulous series of cross-analyses, Wolf identified the key inhibiting factor as close coexistence during the first thirty months of life of either or both of the partners. The

longer and closer the association during this critical period, the stronger
the later effect. Wolf's data allow the reduction or elimination of other
imaginable factors that might have played a role, including the experience of adoption, financial status of the host family, health, age at marriage, sibling rivalry, and the natural aversion to incest that could have
arisen from confusing the pair with true, genetic siblings. Wilson, 175.

195. "Eric Kandel points out": Kandel, 1999, 59.

198. "so many voices in our heads": The "fragmented self" is one of several key
points of congruence between cognitive neuroscience (and evolutionary
psychology) and postmodern cultural theory. (The "decentered," "multi-
plicitous" subject is a fundamental category in the latter tradition, thanks
to the influence of theorists like Jacques Derrida, Gilles Deleuze, and
Julia Kristeva.) Unfortunately, neither the brain scientists nor the cultural
theories seem capable of being in the same room without hurling epithets
at each other. The brain scientists think the cultural theorists are interested only in undermining the empirical claims of science and questioning "truth-claims" at every turn, and the cultural theorists have largely
been uninterested in having brain science shed any light on their theories
of subjectivity. That's a loss for both camps, in my opinion.

198. "repressed wishes": "The UCLA psychologist Robert Bjork and his colleagues have argued persuasively that such directed-forgetting effects are
sometimes attributable to the form of blocking known as retrieval inhibition. Such inhibition can be "released" when we encounter sufficiently
powerful cues that lead us to reexperience an event in the way that we did
initially. Perhaps JR consciously attempted to avoid retrieving memories
of his encounter with the priest and, thus, over a long period of time, successfully inhibited access to them. The potent triggers contained in the
movie may have elicited emotions like those JR felt during the initial
experience, allowing him to overcome the inhibition. Concepts such as
"retrieval inhibition" inevitably call to mind the Freudian notion of
repression. Is retrieval inhibition simply a code word for Freud's old idea,
which has been maligned because it lacks experimental support? Not
really. Freud's concept of repression entails a psychological defense mechanism that is inextricably bound up with attempts to exclude emotionally
threatening material from conscious awareness. But in modern discus-
sions by such theorists as Bjork and Anderson, retrieval inhibition is a far
more ubiquitous construct that applies to both emotional and nonemo-
tional experiences." Schacter, 83.

198. "Darwinian ecosystem": "Perhaps the most vocal contemporary practi-

tioner of neural selectionism is Gerald Edelman, who like Jerne received a Nobel Prize for his work on the immune system. In *Neural Darwinism,* Edelman argued that synapses in the brain, like animals in their environments, compete to stay alive. Synapses that are used compete successfully and survive, while those that are not used perish. According to Edelman, 'The pattern of neural circuitry . . . is neither established nor rearranged instructively in response to external influences.' External influences, instead, select synapses by initiating and reinforcing certain patterns of neural activity that involve them." LeDoux, 2002, 72–73.

203. "neurochemical argument for savoring": "Though researchers have not elucidated the cellular biology of pleasure, theoreticians have given a good deal of thought to anhedonia. The modern reformulation of the concept began in the middle 1970s, when Paul Meehl, a psychologist at the University of Minnesota, published a critique of the prevailing psychoanalytic understanding of hedonic capacity. Meehl stepped back and looked from a distance at the Freudian view of psychological aberration. Freud assumed that all people strive for pleasure, and what distinguishes people are the forces that impede the striving. (To be fair, Freud also thought people differed in the strength of their drives, but that aspect of his thinking was never well elaborated.) The essence of psychoanalysis is the removal of defenses and resistances—various impediments to effective behavior and a full emotional life.

"Meehl considered the impedance of drives to be only half a theory. Yes, people might come to mental illness through fear of various negative consequences; but why should they equally not come to it through the absence of positive reinforcers? His own observation led him to believe that, 'just as there are some organisms impeded by fear, so there are other organisms whose fears are insufficiently softened, attenuated, or, I may even say, impeded by adequate pleasure.' " Kramer, 228–29.

204. "left-brain, right-brain split": "Many a myth has grown up around the brain's asymmetry. The left cerebral hemisphere is supposed to be the coldly logical, verbal, and dominant half of the brain, while the right developed a reputation as the imaginative side, emotional, spatially aware but suppressed. Two personalities in one head, Yin and Yang, hero and villain. To most neuroscientists, of course, these notions are seen as simplistic at best and nonsense at worst. So there was general satisfaction when, a couple of years ago, a simple brain scanner test appeared to reveal the true story about one of neurology's greatest puzzles: exactly what is the difference between the two sides of the human brain? Fortunately, or

unfortunately, depending on how you like your theories, the big picture revealed by that work is proving far less romantic than the logical-creative split, intriguingly complex and tough to prove." McCrone, 2000.

204. " 'triune brain' ": Edelman has a two-part rendition of this model, where the brain stem and the limbic system are considered as a single unit. He also places an important stress on the different speeds of communication prevalent in the two systems: "There are, grossly speaking, two kinds of nervous system organization that are important to understanding how consciousness evolved. These systems are very different in their organization, even though they are both made up of neurons. The first is the brain stem, together with the limbic (hedonic) system, the system concerned with appetite, sexual and consummatory behavior, and evolved defensive behavior patterns. It is a value system; it is extensively connected to many different body organs, the endocrine system, and the autonomic nervous system. Together, these systems regulate heart and respiratory rate, sweating, digestive functions, and the like, as well as bodily cycles related to sleep and sex. It will come as no surprise to learn that the circuits in this limbic–brain stem system are often arranged in loops, that they respond relatively slowly (in periods ranging from seconds to months), and that they do not consist of detailed maps. They have been selected during evolution to match the body, not to match large numbers of unanticipated signals from the outside world. These systems evolved early to take care of bodily functions, they are systems of the interior.

"The second major nervous system organization is quite different. It is called the thalamocortical system. (The thalamus, a central brain structure, consists of many nuclei that connect sensory and other brain signals to the cortex.) The thalamocortical system consists of the thalamus and the cortex acting together, a system that evolved to receive signals from sensory receptor sheets and to give signals to voluntary muscles. It is very fast in its responses (taking from milliseconds to seconds), although its synaptic connections undergo some changes that last a lifetime. As we have seen, its main structure, the cerebral cortex, is arranged in a set of maps, which receive inputs from the outside world via the thalamus. Unlike the limbic–brain stem system, it does not contain loops so much as highly connected layered local structures with massively reentrant connections." Edelman, 1992, 117.

205. "limbic system": LeDoux gives a cautious endorsement to Maclean's model while questioning the accuracy of the limbic system itself in his most recent book, *Synaptic Self.* "Although the limbic system theory is

inadequate as an explanation of the specific brain circuits of emotion, Maclean's original ideas are insightful and quite interesting in the context of a general evolutionary explanation of emotion and the brain. In particular, the notion that emotions involve relatively primitive circuits that are conserved throughout mammalian evolution seems right on target. Further, the argument that cognitive processes might involve other circuits, and might function relatively independent of emotional circuits, at least in some circumstances, also seems correct. These functional ideas are worth preserving, even if we ultimately abandon the limbic system as an anatomical theory of the emotional brain." LeDoux, 2002, 212. Others continue to see the limbic system as a useful category: "To a certain extent, the idea of this conglomerate of regions as the regulator of emotion has been borne out. In many cases, damage to the limbic system results in inappropriate emotion. For example, Klüver-Bucy . . . syndrome occurs when a certain part of the limbic system, the amygdala . . . is damaged. Patients exhibit a high sexual drive, directed not so much toward a prospective partner as toward anything around them, even inanimate objects. Along similar lines, removal of another region, the cingulate cortex . . . in experimental animals results in 'sham' rage—a pattern of behavior that contains all the outward features of a genuine, infuriated state but that occurs for no obvious reason." Greenfield, 4.

206. "damage to their emotional centers": "Now let me submit that a [purely rational decision-making strategy] is not going to work. At best, your decision will take an inordinately long time, far more than acceptable if you are to get anything else done that day. At worst, you may not even end up with a decision at all because you will get lost in the byways of your calculation. Why? Because it will not be easy to hold in memory the many ledgers of losses and gains that you need to consult for your comparisons. The representations of intermediate steps, which you have put on hold and now need to inspect in order to translate them in whatever symbolic form required to proceed with your logical inferences, are simply going to vanish from your memory slate. You will lose track. Attention and working memory have a limited capacity. In the end, if purely rational calculation is how your mind normally operates, you might choose incorrectly and live to regret the error, or simply give up trying, in frustration." Damasio, 1995, 172.

206. "heartbeat to heartstrings to heartless": Wilson, 106.

210. "after her recovery": Freud, 1954, 244.

213. "the shape of South America": Pinker, 2002, 80–81.

BIBLIOGRAPHY

Adolphs, Ralph. "Neural Systems for Recognizing Emotion." *Current Opinion in Neurobiology* (2002).

Amini, Fari, Richard Lannon, and Thomas Lewis. *A General Theory of Love.* New York: Vintage, 2001.

Baron-Cohen, Simon. *Mindblindness: An Essay on Autism and Theory of Mind.* Cambridge, Mass., and London: MIT Press, 1999.

———. "The Extreme Male Theory of The Brain." *TRENDS in Cognitive Sciences* 6, no.6 (June 2002).

Baron-Cohen, Simon, ed. *The Maladapted Mind: Classic Readings in Evolutionary Psychopathology.* East Sussex, U.K.: The Psychology Press, 1997.

Blakemore, Sarah J., Daniel M. Wolpert, and Chris D. Frith. "Central Cancellation of Self-produced Tickle Sensation." *Nature Neuroscience* 1, no. 7. (1998).

Blood, A. J., and R. J. Zatorre. "Intensely Pleasurable Responses to Music Correlate with Activity in Brain Regions Implicated in Reward and Emotion." *Proceedings of the National Academy of Sciences* 98, (2001): 11818–11823.

Blood, A. J., R. J. Zatorre, P. Bermudez, A. C. Evans. "Emotional Responses to Pleasant and Unpleasant Music Correlate with Activity in Paralimbic Regions." *Nature Neuroscience* 2 (2001): 322–27.

Calvin, William. *The Cerebral Code: Thinking a Thought in the Mosaics of the Mind.* Cambridge, Mass., and London, UK: MIT Press, 1996.

Carter, C. S. "Neuroendocrine Perspectives on Social Attachment and Love." *Psychoneuroendocrinology* 23 (1998): 779–818.

Carter, C., A. DeVries, and L. Getz. "Physiological Substrates of Mammalian Monogamy: The Prairie Vole Model." *Neuroscience Biobehavioral Review* 19 (1995): 303–14.

Carter, Rita. *Mapping the Mind.* California: University of California Press, 1998.

Clark, Andy. *Being There: Putting Brain, Body and World Together Again.* London and Cambridge, Mass.: MIT Press, 1997.

Damasio, Antonio. *Descartes' Error: Emotion, Reason, and the Human Brain.* New York: HarperCollins, 1994.

———. *The Feeling of What Happens.* New York: Harcourt, 1999.

Darwin, Charles. *The Expression of the Emotions in Man and Animals.* New York: Oxford University Press, 1998.

Dawkins, Richard. *Climbing Mount Improbable.* New York and London: W.W. Norton, 1996.

———. *The Extended Phenotype: The Long Reach of the Gene.* New York: Oxford University Press, 1982.

———. *Unweaving the Rainbow: Science, Delusion and the Appetite for Wonder.* London: The Penguin Press, 1998.

De Waal, Franz. *Chimpanzee Politics.* Baltimore: Johns Hopkins University Press, 1982.

Dean, Katie. "Attention Kids: Play this Game." *Wired News,* December 19, 2000.

Dehaene, Stanislas, Michel Kerszberg, and Jean-Pierre Changeux. "A Neuronal Model of a Global Workspace in Effortful Cognitive Tasks." *Proceedings of the National Academy of Sciences of the United States of America* 95 (1998): 14529–14534.

Dennett, Daniel C. *Brainchildren: Essays on Designing Minds.* Cambridge, Mass.: MIT Press, 1998.

———. *Consciousness Explained.* Boston, Mass., London, and Toronto: Little, Brown, 1991.

Diamond, Jared. *Why Is Sex Fun?: The Evolution of Human Sexuality.* New York: Basic Books, 1997.

Donaldson, Margaret. *Children's Minds.* New York: W.W. Norton, 1978.

Dreher, J. C., and K. F. Berman. "Fractionating the Neural Substrate of Cognitive Control Processes." *Proceedings of the National Academy of Sciences USA* 99 (2002): 14595–14600.

Edelman, Gerald M. *Bright Air, Brilliant Fire: On the Matter of Mind.* New York: Basic Books, 1992.

———. "Building a Picture of the Brain." *Daedalus* 127 (Spring 1998): 37–69.

———. *Topobiology: An Introduction to Molecular Embryology.* New York: Basic Books, 1988.

Edelman, Gerald, and Giulio Tononi. *A Universe of Consciousness: How Matter Becomes Imagination.* New York: Basic Books, 2000.

Editors of *Scientific American*. *The Scientific American Book of the Brain*. New York: Lyons Press, 1999.

Freud, Sigmund. *Beyond the Pleasure Principle*. trans. James Strachey. New York: W. W. Norton, 1961.

———. "The Uncanny." *The Standard Edition*. Vol. XVII. trans. James Strachey. London: Hogarth Press, 1954.

Gardner, Howard. *Frames of Mind: The Theory of Multiple Intelligences*. New York: Basic Books, 1983.

Greenfield, Susan. *The Private Life of the Brain: Emotions, Consciousness, and the Secret of the Self*. London: John Wiley & Sons, 2000.

Guzeldere, Guven, and Stefano Franchi, eds. *"Bridging the Gap": Where Cognitive Science Meets Literary Criticism*. Stanford Humanities Review Supplement 4, no. 1 (spring 1994).

Hall, Stephen. "Journey to the Center of My Brain." *The New York Times Magazine*, July 1999.

Hauser, Marc D. *Wild Minds: What Animals Really Think*. New York: Henry Holt and Company, 2000.

Hoffman, Donald D. *Visual Intelligence: How We Create What We See*. New York and London: W.W. Norton, 1998.

Hofstadter, Douglas. *Gödel, Escher, Bach: An Eternal Golden Braid*. New York: Basic Books, 1979.

———. *Le Ton beau de Marot: In Praise of the Music of Language*. New York: Basic Books, 1997.

Horgan, John. *Rational Mysticism: Dispatches from the Border Between Science and Spirituality*. New York: Houghton Mifflin, 2003.

Hrdy, Sarah Blaffer, and Sue Carter. "Mothering and Oxytocin or Hormonal Cocktails for Two," *Natural History* (December 1995).

Humphrey, Nicholas. *A History of the Mind: Evolution and the Birth of Human Consciousness*. New York: Springer-Verlag, Copernicus Editions, 1992.

Huxley, Aldous. *The Doors of Perception and Heaven and Hell*. New York: HarperPerennial, 1963.

Insel, T. R., and L. J. Young. "Neurobiology of Social Attachment." *Nature Neuroscience Review* 2 (2001): 129–36.

James, Henry. *The Golden Bowl*. New York: Penguin Classics.

Kandel, Eric. "Biology and the Future of Psychoanalysis: A New Intellectual Framework for Psychiatry Revisited." *American Journal of Psychiatry* (1999), 156:4.

———. "A New Intellectual Framework for Psychiatry." *American Journal of Psychiatry* (1998): 155:457–69.

Kramer, Peter. *Listening to Prozac.* New York and London: Penguin Books, 1993.

LeDoux, Joseph. *The Emotional Brain.* New York: Touchstone, 1996.

———. *Synaptic Self: How Our Brains Become Who We Are.* New York: Penguin Putnam, 2002.

Lumer, E.D., G. M. Edelman, and G. Tononi. "Neural Dynamics in a Model of the Thalamocortical System. I. Layers, Loops and the Emergence of Fast Synchronous Rhythms." *Cerebral Cortex* 7 (1997): 207–27.

McCrone, John. " 'Right Brain' or 'Left Brain': Myth or Reality?" *The New Scientist* (2000).

McGaugh, J. L. "Memory: A Century of Consolidation." *Science,* 2000, 287, 248–51.

McGaugh, J. L., B. Ferry, A. Vazdarjanova, and B. Roozendaal. "Amygdala: Role in Modulation of Memory Storage." *The Amygdala: A Functional Analysis,* J. P. Aggleton, ed. London: Oxford University Press, 2000, 391–423.

McIntosh, Anthony Randal, M. Natasha Rajah, and Nancy J. Lobaugh. "Interactions of Prefrontal Cortex in Relation to Awareness in Sensory Learning." *Science,* 28 May 1999: 1531–1533.

Miller, E. K. "The Prefrontal Cortex and Cognitive Control." *Nature Review of Neuroscience* 1 (2000): 59–65.

Minsky, Marvin. *The Society of Mind.* New York: Touchstone, 1985.

Mithen, Steven. *The Prehistory of Mind: The Cognitive Origins of Art, Religion and Science.* London: Thames and Hudson, 1996.

Ornstein, Robert, and Richard Thompson. *The Amazing Brain.* Boston, Mass.: Houghton Mifflin, 1984.

Panksepp, J., E. Nelson, and M. Bekkedal. "Brain Systems for the Mediation of Social Separation-Distress and Social-Reward: Evolutionary Antecedents and Neuropeptide Intermediaries." *Annals of the New York Academy of Science* 807 (2001): 78–100.

Panksepp, Jaak. *Affective Neuroscience.* New York: Oxford University Press, 1998.

Panksepp, Jaak, and Gunther Bernatzky. "Emotional Sounds and the Brain: The Neuro-Affective Foundations of Musical Appreciation." *Behavioural Processes* 60 (2002): 133–55.

Penrose, Roger. *The Emperor's New Mind: Concerning Computers, Minds, and the Laws of Physics.* New York: Penguin Books, 1991.

Pert, Candace. *Molecules of Emotion.* New York: Simon & Schuster, 1999.

Pinchbeck, Daniel. *Breaking Open the Head: A Psychedelic Journey into the Heart of Contemporary Shamanism.* New York: Broadway Books, 2002.

Philips, William A., and Wolf Singer. "In Search of Common Foundations for Cortical Computation." *Behavioral and Brain Sciences* 20 (1997): 657–722.

Pinker, Steven. *The Blank Slate*. London: Penguin Books, 2002.

———. *The Language Instinct: How the Mind Creates Language*. New York: HarperPerennial, 1994.

Provine, Robert R. *Laughter: A Scientific Investigation*. New York: Penguin Books, 2000.

Quartz, Steven R., and Terrence J. Sejnowski. *Liars, Lovers, and Heroes: What the New Brain Science Reveals About How We Become Who We Are*. William Morrow, New York, 2002.

Restak, Richard. *Brainscapes: An Introduction to What Neuroscience Has Learned About the Structure, Function, and Abilities of the Brain*. New York: Hyperion, 1995.

———. *Mozart's Brain and the Fighter Pilot*. New York: Harmony Books, 2001.

Ridley, Matt. *Genome: The Autobiography of a Species in 23 Chapters*. New York: HarperCollins, 1999.

———. *The Origins of Virtue*, New York: Penguin Putnam, 1996.

Rizzolatti, Giacomo, and Michael Arbib. "Language within Our Grasp." *Trends in Neurosciences*, 21 (1998): 188.

Sacks, Oliver. *An Anthropologist on Mars*. New York: Vintage Books, 1995.

Sagan, Carl, and Ann Druyan. *Shadows of Forgotten Ancestors*. New York: Ballantine Books, 1992.

Schacter, Daniel L. *Searching for Memory: The Brain, the Mind and the Past*. New York: Basic Books, 1997.

———. *The Seven Sins of Memory*. New York: Houghton Mifflin, 2001.

Sime, Wes, Thomas W. Allen, and Catalina Fazzano. "Optimal Functioning in Sport Psychology: Helping Athletes Find Their 'Zone of Excellence.'" *Biofeedback*, Spring 2001.

Stern, D. "The process of therapeutic change involving implicit knowledge: some implications of developmental observations for adult psychotherapy." *Infant Mental Health Journal* 19 (1998): 300–308.

Storr, Anthony. *Music and the Mind*. New York: The Free Press, 1992.

Taylor, John. *The Race for Consciousness*. Cambridge, Mass., and London: MIT Press, 1999.

Taylor, Shelley E. *The Tending Instinct*. New York: Henry Holt and Company, 2001.

Uvnäs-Moberg, Kerstin. "Oxytocin May Mediate the Benefits of Positive Social Interactions and Emotions." *Psychoneuroendocrinology*, 23, no. 8 (1998): 927–44.

Varela, Francisco, Evan Thompson, and Eleanor Rosch. *The Embodied Mind: Cognitive Science and Human Experience.* Cambridge, Mass., and London: MIT Press, 1993.

Vittorio, Gallese, and Alvin Goldman. "Mirror Neurons and the Simulation Theory of Mind-Reading." *Trends in Cognitive Sciences* 2 (1998): 493.

Wilson, Edward O. *Consilience: The Unity of Knowledge.* New York: Random House, 1998.

————. *Sociobiology* (abridged ed.). Cambridge, Mass., and London: Harvard University Press, 1980.

Woolf, Virginia. *Mrs. Dalloway.* New York: Harvest, 1981.

Wright, Robert. *The Moral Animal: Why We Are the Way We Are: The New Science of Evolutionary Psychology.* New York: Random House, 1994.

————. *NonZero: The Logic of Human Destiny.* New York: Pantheon Books, 2000.

ACKNOWLEDGMENTS

This book would have been impossible to write without my many able guides through the world of brain science: Jaak Panksepp, Joseph LeDoux, Shelley Taylor, Sue Carter, Simon Baron-Cohen, Antonio Damasio, John Rodenbough, Wes Sime, Leslie Seiden, Hal Rosenblum, Tom Blue, James McGaugh, Kamran Fallahpour, Susan Othmer, John Donoghue, Robert Provine. I'm grateful to all of them for putting up with my sometimes strange lines of questioning and for collaborating on some of the book's experiments. I'm particularly grateful to Joy Hirsch, who lent me both her fMRI machine and her gifts as a mind reader in making sense of what we discovered.

I'm also indebted to a number of people who read partial or entire drafts of the manuscript and made many helpful suggestions: Simon Baron-Cohen, Joy Hirsch, Antonio Damasio, Aimee Troyen, Gordon Wheeler, John Rodenbough, Eric Liftin, Alexa Robinson, Zack Lynch, and most of all, my research assistant Nesha Burghardt, whose knowledge of the brain sciences and acute eye played an essential role throughout the writing of this book. My friend Maciej Ceglowski created an amazing piece of software for organizing and exploring my research notes. I'd also like to thank the many readers of my website (stevenberlinjohnson.com) who contributed ideas and critiques to the book excerpts I posted online.

Several sections of the book appeared, in modified form, in magazine articles. My wonderful editors at *Discover*—Dave Grogan and Stephen Petranek—

generously let me explore the world of emotions in a three-part series for them. Art Winslow was brave enough to let me publish a defense of evolutionary psychology in *The Nation*. Joel Lovell at *The New York Times Magazine* made many helpful contributions to the sections on neurofeedback. I'd also like to thank Stefanie Syman and Jamie Ryerson for helping me put together the *FEED* special issue on the brain many years ago—my first foray into the world of brain science. I am grateful to the Esalen Center for Theory and Research for inviting me to participate in a conference on Emergence and Consciousness as I was completing this book; the comments and the camaraderie there added a last-minute boost to my work.

Gillian Blake at Scribner took a leap of faith in buying this book originally, given that the premise was very much unproven in the proposal. After Gillian switched houses, Colin Harrison did a masterful job of exposing my stylistic tics and challenging my occasional short-decay ideas, while still inspiring me with his enthusiasm for the book. I apologize to his family for any damage done to his brain by his having to read the manuscript five times. Sarah Knight kept me honest, and more or less on time, throughout the editing process. Nan Graham and Susan Moldow were both inspirational as always. As for my agent, Lydia Wills—it's rare enough to have an agent who does such a good job representing your interests, but it's practically a miracle to have one that also contributes so much intellectually to your publishing career. (Not to mention being fun to talk to on the phone!)

To me, one of the most moving discoveries in the brain sciences—after a century of Darwinian conflict and Oedipal struggle—has been the emerging understanding of the brain's affiliative systems. Our brains are designed to love and connect as much as they are designed to flee and fight. I had ample reminders of those systems every day writing this book: our older boy scrambling up on my lap to pound away on the keyboard; reading copyedits while our one-month-old slept beside me on the couch; debating the fine points of the book's argument with my wife over dinner. I'm grateful to all of them—especially my wife, who knew what she was getting into—for allowing me to draw upon our family experiences at several points in the text. It's strange to think that the boys will someday be old enough to read these pages, but if ten or twenty years from now they should happen to pick up a copy, I hope they'll know how much their presence colored every sentence—and how much I loved these early years together.

New York City
June 2003

INDEX

ABOUT THE AUTHOR

STEVEN JOHNSON is the bestselling author of *Interface Culture, Emergence, Mind Wide Open,* and *Everything Bad Is Good for You.* He is a columnist for *Discover* and a contributing editor at *Wired.* His writing has also appeared in *The New York Times Magazine, The Nation,* and *The Wall Street Journal.* The cofounder of the online magazine *FEED,* he lives in New York City with his wife and two sons. He can be reached via the Web at www.stevenberlinjohnson.com.